Statistics for Anthropology

Second Edition

Anthropology as a discipline is rapidly becoming more quantitative, and anthropology students are now required to develop sophisticated statistical skills. This book provides students of anthropology with a clear, step-by-step guide to univariate statistical methods, demystifying the aspects that are often seen as difficult or impenetrable.

Explaining the central role of statistical methods in anthropology, and using only anthropological examples, the book provides a solid footing in statistical techniques. Beginning with basic descriptive statistics, this new edition also covers more advanced methods such as analyses of frequencies and variance, and simple and multiple regression analysis with dummy and continuous variables. It addresses commonly encountered problems such as small samples and non-normality. Each statistical technique is accompanied by clearly worked examples, and the chapters end with practice problem sets.

Many of the data sets are available for download at www.cambridge.org/9780521147088.

Lorena Madrigal is Professor of Anthropology at the University of South Florida, Tampa. A biological anthropologist, she is particularly interested in the evolution of Afro and Indo Costa Rican populations residing in the Atlantic coast of Costa Rica. She is currently President of the American Association of Physical Anthropologists. She lives in Tampa with her two daughters.

Statistics for Anthropology

Second Edition

LORENA MADRIGAL

University of South Florida, USA

CAMBRIDGE UNIVERSITY PRESS
Cambridge, New York, Melbourne, Madrid, Cape Town,
Singapore, São Paulo, Delhi, Mexico City

Cambridge University Press
The Edinburgh Building, Cambridge CB2 8RU, UK

Published in the United States of America by Cambridge University Press, New York

www.cambridge.org
Information on this title: www.cambridge.org/9780521147088

First published 1998
Second edition 2012

Printed in the United Kingdom at the University Press, Cambridge

A catalogue record for this publication is available from the British Library

Library of Congress Cataloguing in Publication data
Madrigal, Lorena.
Statistics for anthropology / Lorena Madrigal. – Second edition.
pages cm
Includes bibliographical references and index.
ISBN 978-0-521-14708-8
1. Anthropology – Statistical methods. I. Title.
GN34.3.S7M33 2012
301.072′7 – dc23 2011044367

ISBN 978-0-521-14708-8 Paperback

Additional resources for this publication at
www.cambridge.org/9780521147088

This book is dedicated to all parents who never questioned their children's decision to pursue a career in anthropology. To partners who offered support and comfort. And to children who helped us grow to the limit. May all people be as lucky as I have been.

Contents

List of partial statistical tables

Preface

The first edition of this book was published 13 years ago, and much has changed since that time. To put it directly, the first edition of *Statistics for Anthropology* became terribly old-fashioned. When I wrote the first edition it made sense to discuss rounding-off rules, and to dedicate an entire chapter to the computation of frequency distributions. In the year 2012 it does not make any sense to devote time to these topics, as most students will be using a computer package which will easily round the numbers and compute the frequency distributions for them. I have also decided to dedicate very little time to the production of graphs because this is something that college students learned in their pre-college education, and which is easily done with many computer programs. At the same time, I augmented the book by including several issues that were missing in the first edition: the Poisson distribution, two-way ANOVA, the odds ratio, Fisher's exact test, Kendall's tau, and an entire chapter on advanced regression topics. In addition, and to reflect current practice, I have expanded my discussion on probability as to facilitate an understanding of maximum likelihood and Bayesian approaches to statistical inference, as well as to classical hypothesis testing. However, most of the book focuses on the latter, which is still the most frequent approach used in anthropological quantitative analysis. There are specialized books on maximum likelihood and Bayesian approaches which should be consulted for more in-depth discussion on maximum likelihood and Bayesian statistics.

Although this edition is going to be more advanced than the first one, it does not intend to give students their entire statistical education! Instead, this textbook will cover important univariate techniques. Students wishing to learn multivariate statistics need to take a second-level course.

There are two other changes in the book which reflect the passing of time. The first is that no statistical table is included at the end of the book. These tables are now part of the public domain, and they are posted in numerous websites, including this book's website. In my opinion, computer packages have made these tables virtually obsolete because computer packages give analysts the information they used to get in the statistical tables. Therefore, I have only reproduced selected values of statistical tables within the book, enough to illustrate their use as needed in the book's practices and exercises. Another obvious change to this edition is that the data sets used for examples throughout the book are also available in Excel format at the book's website. A description of most of the data sets is included in the website as well. If the data are described within the book, no description is found in the website. When doing the problems by hand I tried to keep as many digits as possible so as to replicate as close as possible the computer output I was getting with SAS or with PASW. However, this does not mean that the answer you will see in the book will be exactly the same (to the very last digit) that you will get with any and all computer packages, many of which use different algorithms. Slight differences in rounding are certainly to be expected.

Most chapters have the same format: (almost) every statistical technique is explained with a mathematical formula and illustrated with an example for which the reader will need a calculator. Afterwards another example will be discussed. Since every instructor is going to choose his or her own computer package, I will not dedicate much time explaining how to work with any particular computer program. Sometimes I will briefly note at the end of the chapter (under computer resources) how to perform the statistical analysis using PASW and SAS, two popular computer packages.

What has not changed since the publication of the first edition is the need for anthropology students to have a solid foundation in statistical analysis. A cursory reading in the major journals in the discipline will show that quantitative methods of data analysis are crucial to anthropological research. When appropriate I will discuss research articles which use the statistical technique I am discussing. I would like to note that this book covers statistical techniques to analyze quantitative data. Therefore, I do not cover any kind of qualitative data analysis.

I wish to thank the instructors who adopted the first edition, and the students who told me that they enjoyed learning statistics with the first edition. I also want to thank everyone who gave me suggestions for improving the book, whether personally or in writing. In this second edition I endeavored to remove all typos and errors which did exist in the first one. However, as a typical member of the *Homo sapiens* species I may have made a few mistakes. Please contact me at madrigal@usf.edu to let me know of any.

Happy computing!

1 Introduction to statistics and simple descriptive statistics

This chapter discusses several topics, from why statistics is important in anthropological research to statistical notation. The first section (statistics and scientific enquiry) defines basic scientific terms and explains the role of statistics in anthropological research. The second section (basic definitions) reviews the vocabulary we need for the rest of the book. The third section (statistical notation) explains the fundamentals of statistical notation.

1.1 Statistics and scientific enquiry

If you are an anthropologist, you have probably been asked what anthropology is, and what it is good for. Many of us are at a loss to explain the obvious: how else could we look at the world, but with a cross-cultural and evolutionary perspective? What you may not be quite convinced about is the need for you to include a statistical aspect to your anthropological data analysis. In this section, I hope to explain why statistics are an integral part of a scientific approach to anthropological enquiry.

The word **statistics** is part of popular culture and every-day jargon. For that reason, I wish to clarify our working definition of the term in this book. Let us agree that statistics are the figures which summarize and describe a data set (**descriptive statistics**), and the methods used to arrive at those figures. In addition, statistical analysis allows us to make predictions about the wider universe from which a data set was obtained (**inferential statistics**). The reason statistics are an integral part of the anthropology curriculum is that statistical methods allow us to approach our subject of study in a manner which lets us test hypotheses about the subject. Or as G. Jenkins says: "The preoccupation of some statisticians with mathematical problems of dubious relevance to real-world problems has been based on the mistaken notion that statistics is a branch of mathematics – In contrast to the more sensible notion that it is part of the mainstream of the methodology of science" (Jenkins 1979). Indeed, some students enter my classroom fearful because they say they are not good at mathematics. However, the statistics class offered by my department is not a mathematics class; it is a class on methodology useful for testing hypotheses in anthropology.

There are other terms (such as hypothesis) which are used with different meanings in popular and scientific parlance. Within science, different disciplines have different definitions for them. For that reason I wish to define some key terms at the start of the book.

I would first like to define the term **fact**. It is a fact that you are reading this book. Such fact is easily verified by yourself and others. People who are visually impaired can verify this by using their sense of touch and by asking other people to confirm this. Therefore, a fact is something that is verified usually, though not always by human senses. The existence of sub-atomic particles is verified by computers. Although a human might need her senses to read or feel the output generated by the computers, it is the latter which was able to detect the sub-atomic particles. Thus, verification of facts is not limited by human senses (otherwise, how could we detect sound used by other animal species which we humans are unable to detect?). Science would have hardly advanced if we had limited observation to what we humans are able to detect with our senses. Thus, facts are verifiable truths, where the verification is not limited to human senses.

A **hypothesis** is an explanation of facts. What makes a hypothesis scientific is that it can be tested and rejected by empirical evidence. A hypothesis that cannot be tested is not scientific. Scientific hypotheses explain observed facts in testable ways. For example, it is a well-known fact that the frequency of different colors changed in peppered moths in London during the height of the industrial revolution. Prior to the onset of heavy pollution in London, the majority of the moth population was lightly pigmented because light-colored moths were well camouflaged on light tree bark from their predators (birds). However, due to heavy pollution, tree bark became progressively darker. As a result, dark moths became better camouflaged, and the frequency of light moths decreased and dark ones increased. The frequency of dark peppered moths increased from less than 1% in 1848 to 95% in 1898. With the control of pollution in the 1900s, tree barks became lighter, light moths had the survival advantage, and the frequency of dark moths decreased. These are the observed and verifiable facts. Various hypotheses can be proposed to explain these facts. For example, it could be proposed that such changes in moth coloration are the result of a supernatural Being testing our faith. Or it could be hypothesized that birds acted as a natural selection agent, and that the change in moth coloration was an evolutionary change experienced by the moth population. Both propositions are explanations of the facts, but only the second one can be empirically tested and therefore be considered scientific. Please note that a hypothesis cannot be proven to be true. It can, however, be rejected. A hypothesis that has not been rejected after many studies is more likely to be correct than one that has been supported by only a single study or none at all.

There is quite a difference in the meaning of **theory** in popular culture and in science. In the former, the word theory is sometimes used dismissively, as if it were something with no factual base. This is certainly not how a theory is understood to be in science. We define a **theory** as a set of unified hypotheses, none of which has been rejected. For example, the theory of plate tectonics encompasses several hypotheses which explain several facts: the shape of Africa and South America "fit," the stratigraphy of both continents also correspond with each other, the shape of Madagascar "fits" with Africa, etc. The currently accepted (not proven) hypotheses to explain these facts is that there is continental movement, that Africa and South America were once a single land mass, and that Madagascar split from Africa. Therefore, a theory is able to explain facts with hypotheses driven by the theory. These hypotheses are tested and accepted (for the moment). Should any of these hypotheses be rejected later (perhaps because better observation is possible as a result of new equipment), then the theory encompassing the rejected hypothesis must be revisited. But the entire theory does not fall apart.

The example above illustrates what is (in my opinion) the most distinctive trait of science as a form of human knowledge different from other forms of knowledge: science is by definition a changing field. Hypotheses which have been accepted for decades could very well be rejected any day, and the theory which drove those hypotheses revisited. As Futuyma so clearly puts it: "... good scientists *never* say they have found absolute 'truth' (emphasis in text) (Futuyma 1995).

Statistical methods are of fundamental importance for the testing of hypotheses in science. Researchers need an objective and widely recognized method to decide if a hypothesis should be accepted or rejected. Statistical tests give us this method. Otherwise, if each of us were to decide on what criteria to use to accept or reject hypotheses, we would probably never allow ourselves the opportunity to accept a hypothesis we want to reject or vice versa. As a tool to test hypotheses and advance theories, statistics are an integral part of scientific studies.

There is one more reason why statistics are so important for hypothesis-driven anthropological research: if we quantify results, we are able to compare them. The need for comparing results is due to the fact that scientific results should be replicable. As you must have learned in high school science classes, different people following the exact same procedures in a scientific experiment should be able to obtain the same results. The problem in anthropology, of course, is that it is impossible to replicate a historical event such as a migration. But if we use statistics to summarize data, we can compare our results with the results of other researchers. For example, I am interested in determining if migrant communities have increased body mass index (BMI, computed as $BMI = \frac{kg}{m^2}$) and hypertension rates in comparison with non-migrant communities. Although I cannot replicate a migration event, I can compare data on BMI and hypertension in several migrant communities and determine if the migrant communities do or do not differ from non-migrant communities. By using statistical methods, we were able to show that whereas migrant groups experienced an increase in BMI, they did not always experience an increase in hypertension rates (Madrigal *et al.* 2011). We should remember that anthropology is by definition a comparative science. A cross-cultural view of anything human is intrinsic to the field. With statistics anthropologists are able to compare their results with the results of others.

1.2 Basic definitions

1.2.1 Variables and constants

One sure way to favorably impress your instructor is to refer to data in the plural and not singular. You can say, for example, that your data are a collection of measures on a group of 100 women from a village. Your data include the following information: woman's age, religion, height and weight and number of children produced. The **unit of analysis** in this data set is the woman, from each of whom you obtained the above information, which includes both variables and constants. The fact that all of your subjects are women and the fact that they all live in the same village means that gender and village are both constants. **Constants** are observations recorded on the subjects which do not vary in the sample. In contrast, **variables** are observations which do vary from subject to subject. In this example the number of children produced, the age, the height, and the weight all vary. They are therefore variables. A singular observation in a subject (age 25, for example) may be referred to as an **observation**, an individual **variate** or as the **datum** recorded in the subject.

1.2.2 Scales of measurement

A cursory look at statistics textbooks will indicate that different authors favor different terms to refer to the same concept regarding the scale of measure of different variables. Please do not be surprised if the terms I use here are different from those you learned before.

1.2.2.1 Qualitative variables

Qualitative variables classify subjects according to the kind or quality of their attributes. These variables are also referred to as attributes, categorical or nominal variables. An example of such variables is the religion affiliation of the women. If an investigator works with qualitative variables, he may code the different variates with numbers. For example, he could assign a number 1 to the first religion, 2 to the second, etc. However, simply because the data have been coded with numbers, they cannot be analyzed with just any statistical method. For example, it is possible to report the most frequent religion in our sample, but it is not possible to compute the mean religion. I have always preferred to enter qualitative data into spreadsheets by typing the **characters** (Christian, Muslim, etc.) instead of using numbers as codes. However, not all computer packages allow you to enter data in this manner.

Another important point about research with qualitative variables concerns the coding system, which should consist of mutually exclusive and exhaustive categories. Thus, each observation should be placed in one and only one category (**mutual exclusiveness**), and all observations should be categorized (**exhaustiveness**). Qualitative variables will help us group subjects so that we can find out if the groups differ in another variable. For example, we could ask if the women divided by religion differ significantly in the number of children they produced, or in their height or weight. Qualitative variables themselves will be the focus of our analysis when we ask questions about their frequency in chapter eight.

1.2.2.2 Ranked or ordered variables

Ranked or ordered variables are those whose observations can be ordered from a lower rank to a higher rank. However, the distance or interval between the observations is not fixed or set. For example, we can rank individuals who finish a race from first to last, but we do not imply that the difference between the first and second arrivals is the same as that between the second and the third. It is possible that the difference between the first and second place is only 2 seconds, while the difference between the second and third place is 3 minutes. A more appropriate anthropological example would be a situation in which we do not have the actual age of the women we interviewed because they do not use the Western calendar. In this case, we could still rank the women from youngest to oldest following certain biological measures and interviews to confirm who was born before whom. We will use ranked variables quite a bit in our non-parametric tests chapter (chapter seven).

1.2.2.3 Numeric or quantitative variables

Numeric or quantitative variables measure the magnitude or quantity of the subjects' attributes. Numeric variables are usually divided into discontinuous/discrete and continuous variables.

Discontinuous numerical variables have discrete values, with no intermediate values between them, while the distance between any two values is the same (as opposed to ranked variables). In the research project mentioned above, the number of children born to a woman is an example of discontinuous variables because it can only be whole numbers. Also, the difference between one and two children is the same as the difference between 11 and 12 children: the difference is one child. As we well know, discontinuous numeric variables are amenable to statistical analysis which may produce counterintuitive results. If we compute the mean number of children of two women, one of whom produced one and one of whom produced two children, the result will be 1.5 children. Therefore, it is possible to compute the mean of discrete numerical variables, whereas it is not possible to compute the mean of qualitative variables such as religious membership, *even if the latter are coded with numbers*.

Continuous numeric variables are numeric data which do allow (at least theoretically) an infinite number of values between two data points. In the research project mentioned above, the weight and height of the women are continuous numeric variables. In practice, investigators working with continuous variables assign observations to an interval which contains several measurements. For example, if an anthropologist is measuring her subjects' height, and a subject measures 156.113 cm and another measures 155.995, the researcher will probably assign both subjects to one category, namely, 156 cm. White (1991) discusses the issue of measurement precision in osteological research. He specifically focuses on the appropriate procedure to follow when slightly different measures of the same tooth or bone are obtained. The problems associated with measurement in osteology show that the measurement of continuous variables is approximate, and that the true value of a variate may be unknowable (White 1991). You will sometimes see a distinction between **interval** and **ratio** continuous numeric variables. In an interval scale a value of 0 does not mean total absence of the item measured by the scale. For example, a 0 value for temperature measured in the Fahrenheit or Celsius scales does not mean absence of temperature. In contrast, a 0 value in a ratio variable does mean total absence, such as 0 kilos. In terms of statistical manipulation, the difference between ratio and interval scales is not important, so they are both treated in the same manner in this book. The analysis of continuous numeric data is the main purpose of this book.

1.2.3 Accuracy and precision

An **accurate measurement** is one that is close to the true value of that which is measured. When doing research, we should strive to obtain accurate data, but this is not as easy as it sounds for some variables. If we are working with easily observable discrete numeric variables (let's say the number of people in a household at the time when we visit it) then it's easy to say that there are three, six, or ten people. If the variable we wish to measure is not so easily observable (let's say the number of children produced by the woman) we might not be able to determine its true value accurately. It is possible that the woman had a baby when she was young and gave it up for adoption, and nobody in her household knows about it. She is not going to tell you about this baby when you interview her. In this case the accurate (true) number of children produced by this woman is the number of children she declares plus one (assuming that the number of children she declares is accurate). The problems associated with obtaining accurate measures of continuous numeric variables are different, and I already alluded to them in section 1.2.2.3. The better the instrument for measuring height in living subjects, or length of a bone, the more accurate the measurement. If we can determine with a tape measure that a subject's height is 156 cm but with a laser beam that she is 156.000789 cm, the latter measure is more accurate than the former. A **precise measure** is one that yields consistent results. Thus, if we obtain the same value while measuring the height of our subject, then our measure is precise. Although a non-precise measure is obviously non-accurate, a precise measure may not be accurate. For example, if you interview the woman about how many children she had, she may consciously give you the same response, knowing that she is concealing from you and her family that one baby she had when she was very young and gave up for adoption. Since she gives you the same response, the answer is precise; but it is not accurate.

1.2.4 Independent and dependent variables

The independent variable is the variable that is manipulated by, or is under the control of the researcher. The **independent variable** is said to be under the control of the experimenter because she can set it a different level. The **dependent variable** is the one of interest to the researcher, and it is not manipulated. Instead, she wishes to see how the independent variable affects the dependent variable, but she does not interfere or manipulate the latter. In a laboratory setting, it is easier to manipulate an independent variable to see its effects on the dependent one. Many readers have seen films of the Harry Harlow experiments on the effects of isolation on the behavior of young monkeys, in which the independent variable was degree of isolation, and the dependent variable was the behavior of the animals. For example, Harlow raised some monkeys with their mothers, while he raised others in the company of other young monkeys and he raised others alone. By varying the degree of isolation, Harlow manipulated the independent variable, and observed its effects on the behavior of the monkeys.

In a non-laboratory setting it is much more difficult to have such tight control over an independent variable. However, according to the definition above, the independent variable is under the control of the researcher. Thus, we can separate the women in our research project by religion (independent variable) and ask if the two groups of women differ in their mean number of children (dependent variable).

In this book we will denote independent variables by an X and dependent variables by a Y. This distinction will be important in our regression chapters only. However, since it is our wish to understand the behavior of the dependent variable, we will usually refer to a variable with the letter Y. If we are discussing more than one variable we will use other letters, such as X, Z, etc.

1.2.5 Control and experimental groups

Let us go back to the research project in which you have data on a group of women from a village, from whom you collected each woman's age, religion, number of children produced, height, and weight. Let us say that you are working for an NGO which seeks to give some kind of employment and therefore better economic prospects to the women. Let us say that you recruit 50 women into the new program for an entire year and keep 50 out of the program. The following year you measure the 50 women in the program and the 50 not in the program for all the same variables. It would be more proper to refer to these two groups as the experimental group (the women in the program) and the control group (the women not in the program). Therefore, the **experimental group** receives a **treatment**, while the **control group** remains undisturbed, and serves as a comparison point. Assuming that participation in this program is beneficial to the women, we could predict that we would see a difference in the two groups of women. Specifically, we could predict that women in the program will have a healthier weight after a year when compared with women not in the program. It is by having a control group that we are able to show that a change does or does not have an effect on our subjects. Please note that we could express this example using independent/dependent variables terminology: in this example the independent variable is participation in the program (yes or no, under the control of the investigator) and the dependent variable is weight (which we do not manipulate).

If subjects are to be divided into experimental and control groups, the statistical decision derived from the experiment rests on the assumption that the assignment to groups was done randomly. That is, the researcher must be assured that no uncontrolled factors are influencing the results of the statistical test. For example, if you assign to the new program only women of one religion, and use as a control group the women of the other religion, and the mean weight between the groups differs, you do not know if you are seeing the effects of religion, the program, or both combined, on weight. If subjects are randomly assigned to treatment or control groups and the treatment does not have an effect, then the results of the experiment will be determined entirely by chance and not by the treatment (Fisher 1993).

1.2.6 Samples and statistics, populations and parameters. Descriptive and inferential statistics. A few words about sampling

A statistical population is the entire group of individuals the researcher wants to study. Although statistical populations can be finite (all living children age 7 in one particular day) or infinite (all human beings when they were 7 years old), they tend to be incompletely observable (how could all children age 7 in the world be studied in one day?). A **parameter** is a measure (such as the mean) that characterizes a population, and is denoted with Greek letters (for example, the population mean and standard deviation are designated with the Greek letters μ *-mu-* and σ *-sigma-* respectively). But since populations are usually incompletely observable, the value of the population parameter is usually unknown.

A **sample** is a subset of the population, and generally provides the data for research. Some students upon taking their first statistics class feel that there is something wrong about the fact that we don't work with populations but rather with samples. Please do not let this bother you. Most research in realistic situations must take place with samples. You should however be concerned with obtaining a representative sample (this is discussed below). A **statistic** is a measure that characterizes a sample. Thus, if a sample of children age 7 is obtained, its average height could easily be computed. Statistics are designated with Latin letters, such as $\bar{Y}$ (Y-bar) for the sample mean and s for the sample standard deviation. This difference in notation is very important because it provides clear information as to how the mean or standard deviation were obtained. It should also be noted that population size is denoted with an uppercase N, whereas sample size is denoted with a low case n. In this book we will differentiate the **parametric** from the **sample notation**.

Descriptive statistics describe the sample by summarizing raw data. They include measures of central tendency (the value around which much of the sample is distributed) and dispersion (how the sample is distributed around the central tendency value) such as the sample mean and standard deviation respectively. Descriptive statistics are of extreme importance whether or not a research project lends itself to more complex statistical manipulations.

Inferential statistics are statistical techniques which use sample data, but make inferences about the population from which the sample was drawn. Most of this book is devoted to inferential statistics. Describing a sample is of essential importance, but scientists are interested in making statements about the entire population. Inferential statistics do precisely this.

Sampling. Since this book is about statistical analysis and not about research design, I will not discuss the different types of sampling procedures available to researchers. Moreover, the research design and sampling of anthropologists can vary a lot, whether they are doing paleoanthropology, primatology, door-to-door interviews, or archaeological excavations. However, I would like to discuss two issues.

(A) Samples must be representative and obtained with a random procedure. In the research project we have been discussing in this chapter, our data set included women from both religious groups. This was done because if we had only measured women of one group, our village sample would have been biased, or it would not accurately represent the entire village. A **representative sample** is usually defined as having been obtained through a procedure which gave every member of the population an equal chance of being sampled. This may be easier said than done in anthropology. An anthropologist in a particular community needs to understand the nuances and culture of the population, to make sure that an equal chance of being sampled was given to each and every member of the population. In many instances, *common sense is the most important ingredient to a good sampling procedure*. If you know that the two religious groups are segregated by geography, you need to obtain your sample in both areas of the village so that you have members of both groups (sample is representative).

There are some situations in anthropological research in which random sampling can hardly be attempted. For example, paleoanthropologists investigating populations of early hominids would hope to have a random sample of the entire population. But these researchers can only work with the animals that were fossilized. There is really no sampling procedure which could help them obtain a more representative sample than the existing fossil record. In this situation, the data are analyzed with the acknowledgment that they were obtained through a sampling procedure that cannot be known to be random.

(B) Samples must be of adequate size. The larger the sample, the more similar it is to the entire population. But what exactly is large? This is not an easy question, especially because in anthropology it is sometimes impossible to increase a sample size. Paleoanthropologists keep hoping that more early hominids will be unearthed, but can only work with what already exists. However, if a research project involves more easily accessible data sets, you should consider that most statistical tests work well (are **robust**) with samples of at least 30 individuals. Indeed, there is a whole suite of non-parametric statistical tests specifically designed for (among other situations) cases in which the sample size is small (discussed in chapter seven). As I mentioned above, sometimes the most important aspect of research design is common sense. When you are designing your project and you are trying to determine your ideal sample size, you should talk to experts in the field and consult the literature to determine what previous researchers have done. In addition, you might be able to perform a power analysis to help you determine your ideal sample size, although not everyone has the necessary information to do this. Power analysis is discussed in chapter four.

1.3 Statistical notation

Variables are denoted with capital letters such as X, Y, and Z while individual variates will be denoted with lower case letters such as x, y, and z. If more than one variable is measured in one individual, then we will differentiate the variables by using different letters. For example, we might refer to height as Y, and to weight as X. Distinct observations can be differentiated through the use of subscripts. For example, y_1 is the observation recorded in the first individual, y_2 is the observation recorded in the second one, and y_n is the last observation, where n is the sample size.

Sigma ($\sum$) is a **summation sign** which stands for the sum of the values that immediately follow it. For example, if Y stands for the variable height, and the sum of all the individuals' heights is desired, the operation can be denoted by writing $\sum Y$. If only certain values should be added, say, the first ten, an index in the lower part of sigma indicates the value at which summation will start, and an index in the upper part of sigma indicates where summation will end. Thus: $\sum_1^{10} Y = y_1 + y_2 + \ldots + y_{10}$. If it is clear that the summation is across all observations, no subscripts are needed. Below are two brief examples of the use of sigma:

X	Y
6	7
8	9
5	2
3	10
9	1
10	3
$\sum X = 41$	$\sum Y = 32$

A frequently used statistic is $(\sum Y)^2$ which is the sum of the numbers, squared. In our example: $(\sum X)^2 = 41 = 1681$ and $(\sum Y)^2 = 32 = 1024$.

Another frequently used statistic is $\sum Y^2$ which is the sum of the squared numbers (and is called the uncorrected sums of squares by SAS). Using our previous examples, we can square the numbers, and sum them as follows:

X	X^2	Y	Y^2
6	36	7	49
8	64	9	81
5	25	2	4
3	9	10	100
9	81	1	1
10	100	3	9
$\sum X^2 = 315$		$\sum Y^2 = 244$	

The reader should not confuse $\sum Y^2$ with $(\sum Y)^2$. The former refers to the sum of squared numbers, whereas the latter refers to the square of the sum of the numbers. These two quantities are used in virtually every statistical test covered in this book.

If the sample includes two columns of numbers, it is possible to obtain additional statistics. If a column is multiplied by another, then a third column is obtained, which is the product of both columns. The sum of that third column is denoted by $\sum XY$. This is illustrated with the previous data:

X	Y	XY
6	7	42 (6*7)
8	9	72 (8*9)
5	2	10 (5*2)
3	10	30 (3*10)
9	1	9 (9*1)
10	3	30 (10*3)
		$\sum XY = 193$. $\left(\sum XY\right)^2 = 37{,}249$

Practice problem 1.1

Using these two columns of data, obtain the following statistics:

$$\sum X, \left(\sum X\right)^2, \sum Y, \sum Y^2, \left(\sum XY\right), \left(\sum XY\right)^2$$

X	Y
0	15
70	80
100	6
10	50
20	75

Let us do the calculations step by step:

$$\sum X = 0 + 70 + 100 + 10 + 20 = 200$$

$$\left(\sum X\right)^2 = (200)^2 = 40{,}000$$

$$\sum X^2 = (0)^2 + (70)^2 + (100)^2 + (10)^2 + (20)^2$$
$$= 0 + 4{,}900 + 10{,}000 + 100 + 400 = 15{,}400$$

$$\sum Y = 15 + 80 + 6 + 50 + 75 = 226$$

$$\sum Y^2 = (15)^2 + (80)^2 + (6)^2 + (50)^2 + (75)^2$$
$$= 225 + 6{,}400 + 36 + 2500 + 5625 = 14{,}786$$

$$\left(\sum Y\right)^2 = (226)^2 = 51{,}076$$

$$\sum XY = (0)(15) + (70)(80) + (100)(6) + (10)(50) + (20)(75)$$
$$= 0 + 5600 + 600 + 500 + 1500 = 8200$$

$$\left(\sum XY\right)^2 = (8{,}200)^2 = 67{,}240{,}000$$

1.4 Chapter 1 key concepts

- Statistics.
- Difference between science and other forms of knowledge.
- Scientific hypothesis.
- Theory.
- Science and repeatability.
- The role of statistics in science.
- Variables and constants.
- Scales of measurement.
- Accuracy and precision.
- Independent, dependent variables.
- Control and experimental groups.
- Samples and statistics, populations and parameters.
- Problems with sampling: randomness and size.
- Descriptive and inferential statistics.
- Statistical notation.

1.5 Chapter 1 exercises

All data sets are available for download at www.cambridge.org/9780521147088.

1. Write down a research project in your own area of study. State the following:
   a. Describe the variables you will collect. What is the scale of these variables?
   b. Which variables are dependent and which independent?
   c. Are you going to have control and experimental groups?
   d. How are you going to obtain your sample?
   e. How are you going to assure accuracy and precision in your data?
2. Use the data set called "family size." Let education = Y and income = X. Compute

$$\sum X, \left(\sum X\right)^2, \sum Y, \sum Y^2, \left(\sum XY\right), \left(\sum XY\right)^2$$

3. Use the "menarche" data set. Let age at menarche = Y and age at first pregnancy = X. Compute

$$\sum X, \left(\sum X\right)^2, \sum Y, \sum Y^2, \left(\sum XY\right), \left(\sum XY\right)^2$$

2 The first step in data analysis: summarizing and displaying data. Computing descriptive statistics

In this chapter we review what should be the first steps in data analysis. A sure way to aggravate your adviser would be to bring her a table with your raw data (hundreds of observations) for her reading pleasure. Instead of making such a mistake, you would do well to summarize your data into an easy-to-read table, namely, a frequency distribution, and to illustrate this table with a graph. Afterwards, you should compute simple descriptive statistics that summarize the sample. This chapter is divided into these two broad sections: (1) frequency distributions and graphs and (2) descriptive statistics.

2.1 Frequency distributions

2.1.1 Frequency distributions of discontinuous numeric and qualitative variables

Frequency distributions are very useful as a first step to understand how data are distributed, that is, what values are most frequent in a sample, and which ones are less frequent or even absent. The procedure for constructing a frequency distribution of a discontinuous numeric variable and a qualitative variable is very similar, so I discuss the procedure for both variables together. Tables 2.1 and 2.2 show a frequency distribution of a discontinuous numeric variable (number of children produced) and a qualitative variable (religious membership). The latter one is rather uninteresting because there are only two categories (religious group one and two), so we discuss the former one.

A frequency distribution should have as a first column a listing of the observations measured in the data set. You will notice that in group one we observed variates ranging from one through fourteen, but that variate thirteen was not recorded, so it is missing in the table. In contrast, the listing of observations for group two starts with the variate five and finishes with fifteen. Therefore, in a frequency distribution it does not make sense to list an observation (13 in group one or 1–4 in group two) which was not observed. The second column should list the frequency at which an observation was recorded. You will notice that whereas four women in both groups produced five children, two women in group one and seven in group two produced 12 children. The third column shows this information in a percent format. Since the sample size of both groups is 50, the percent of five children in both groups is $\frac{4}{50} = 8\%$, while the percent of 12 children

Table 2.1 Frequency distribution of number of children produced by women who belong to two religious groups.

Number of children produced	Frequency	Percent	Cumulative frequency	Cumulative percent
Group 1				
1	1	2.00	1	2.00
2	1	2.00	2	4.00
3	3	6.00	5	10.00
4	4	8.00	9	18.00
5	4	8.00	13	26.00
6	7	14.00	20	40.00
7	5	10.00	25	50.00
8	8	16.00	33	66.00
9	7	14.00	40	80.00
10	3	6.00	43	86.00
11	4	8.00	47	94.00
12	2	4.00	49	98.00
14	1	2.00	50	100.00
Group 2				
5	4	8.00	4	8.00
6	2	4.00	6	12.00
7	3	6.00	9	18.00
8	7	14.00	16	32.00
9	11	22.00	27	54.00
10	5	10.00	32	64.00
11	4	8.00	36	72.00
12	7	14.00	43	86.00
13	4	8.00	47	94.00
14	2	4.00	49	98.00
15	1	2.00	50	100.00

Table 2.2 Frequency distribution of religious affiliation in a sample of 100 women.

		Frequency	Percent	Cumulative frequency	Cumulative percent
Group	1	50	50.0	50.0	50.0
	2	50	50.0	100.0	100.0
	Total	100	100.0		

in the first group is $\frac{2}{50} = 4\%$ and in the second group is $\frac{7}{50} = 14\%$. The percent column gives the reader important information: a family size of 12 is much more frequent in the second rather than the first group. The last two columns of a frequency distribution give you similar information, but in frequency and percent formats. How many women produced five or fewer children in groups one and two? To answer this question we look at the cumulative frequency column: while 13 women produced five or fewer children in the first group, only four women did so in group two. In the same way, we can say that 26% of all women in the first group produced five or fewer children, while only 8% of all women did so in the second group. You can see that the last two columns provide very rich information about how the data are distributed in our two groups. The frequency distribution of both groups indicates that women in group one tend to produce fewer children than do women in group two.

2.1.2 Frequency distributions of continuous numeric variables

Constructing a frequency distribution of discontinuous numerical and qualitative data is easy enough because it is easy to classify subjects since they have an identifiable observed variate: a woman has zero, one, two, or three living children. Even if a woman is not sure about how many surviving children she currently has, you can classify her into a category such as "indeterminate value." In contrast, constructing a frequency distribution of continuous numerical data is more complicated due to the precision involved in the measurement of the data. The problem of doing frequency distributions with continuous data arises precisely during the process of assigning subjects to categories. For example, if you measured two subjects' heights as 156.00023 and 155.999 cm you will probably assign them to the same category: 156. Otherwise, if you decide to keep each digit above, chances are that each subject will end up being assigned to her own category. Table 2.3 shows very precise measures of height for ten subjects. When a frequency distribution was requested from both PASW and SAS, both gave me a frequency distribution with ten categories, one for each subject. Therefore, no economy was achieved. It is clear that we need to group the subjects into intervals, and report the frequency and percent of the interval.

Frequency distributions of continuous data are useful not only as a first stage when summarizing the data set, but also as an intermediate stage in the hand-computation of statistics with such data. If you ever find yourself without a computer program which computes statistics, and you have a significantly large data set, you would need to arrange your continuous data into frequency distributions to facilitate such computations. In the first edition of this book, I devoted quite a few pages to how to compute a frequency distribution of continuous data. In this edition, I decided against doing so because the need to construct frequency distributions by hand of such data has virtually disappeared. I cannot imagine that any reader of this edition will ever be without a computer program which computes statistics. Most anthropology students are in the field or in the lab entering data (from hormonal analysis of urine samples to display behavior in non-human primates to number of projectile points in an archaeological site) with a computer next to them. Once they enter the data into the computer, these students will proceed to compute statistics with the same computer. Therefore, the need to explain how to compute a frequency distribution of continuous data as an intermediate stage in the computation of statistics, in my opinion, has disappeared.

Table 2.3 Raw data and frequency distribution of height in ten subjects (ungrouped data).

Raw data	
Observation	Height
1	179.58065582
2	156.68456038
3	161.43566923
4	156.49174451
5	138.00431500
6	164.41657784
7	170.25923455
8	163.65883671
9	157.44467354
10	156.41354636

Frequency distribution

Height	Frequency	Percent	Cumulative frequency	Cumulative percent
138.00431500	1	10.00	1	10.00
156.41354636	1	10.00	2	20.00
156.49174451	1	10.00	3	30.00
156.68456038	1	10.00	4	40.00
157.44467354	1	10.00	5	50.00
161.43566923	1	10.00	6	60.00
163.65883671	1	10.00	7	70.00
164.41657784	1	10.00	8	80.00
170.25923455	1	10.00	9	90.00
179.58065582	1	10.00	10	100.00

Table 2.4 A frequency distribution of height in 100 subjects.

Height	Frequency	Percent	Cumulative frequency	Cumulative percent
123–132	1	1.00	1	1.00
133–142	5	5.00	6	6.00
143–152	17	17.00	23	23.00
153–162	39	39.00	62	62.00
163–172	29	29.00	91	91.00
173–182	8	8.00	99	99.00
183–192	1	1.00	100	100.00

If you do want to compute a frequency distribution using a computer program to display your data, you will need to tell the program how to group the subjects. Therefore, you need to declare that you want subjects grouped in five-year or ten-year categories, for example. You should make sure, of course, that you check that every subject has been assigned to one category. For example, when I constructed the frequency distribution

Table 2.5 Stem-and-leaf display for height of 100 women divided by religious affiliation.

Religion = 1	
Frequency	Stem & Leaf
3.00	13.599
2.00	14.24
4.00	14.6779
9.00	15.011233334
8.00	15.66778899
9.00	16.011123334
7.00	16.5557889
6.00	17.011234
2.00	17.56
Stem width:	10
Each leaf:	1 case(s)
Religion = 2	
Frequency Stem & Leaf	
4.00	14.3334
2.00	14.56
10.00	15.0012344444
14.00	15.55555666777899
6.00	16.013444
9.00	16.555666889
4.00	17.0123
1.00	17.5
Stem width:	10
Each leaf:	1 case(s)

of height shown in Table 2.4, I had mistakenly left the highest value out. When SAS printed the frequency distribution, it told me that there was one missing observation, which alerted me to my mistake.

Please note that the columns in Table 2.4 are the same as those of Tables 2.1 and 2.2. The table informs us that 39% of the women are between 153 and 162 cm tall, and that 62% of the women are 153–162 or less cm tall.

2.1.3 Stem-and-leaf displays of data

This type of data summary or display is the answer to the question: "What do you get when you cross a frequency distribution and a graph?" Indeed, whereas some authors prefer to discuss it under the "graphs" section (and PASW refers to it as a plot) others do so under the "frequency distribution" section. I chose the latter approach because in my opinion, a stem-and-leaf display is most useful as a tool to summarize data rather than to graphically display it. I particularly favor it for summarizing continuous numerical data. This type of display features a column with the frequency of observations, where the observations are grouped in a manner that makes sense in terms of the precision with which variables were measured. Table 2.5 shows a stem-and-leaf display of height rounded to the nearest centimeter (170 cm, 180 cm, etc.) for the women in our sample, divided by religious affiliation. Let us look at the data for the first religious group. The first column (frequency) tells you how frequent observations in the range shown in the second column were (3), while the second column tells you that you are dealing with women between 130 and 140 cm tall. As you move to the right of 13, you will see the following numbers: 599. This information is interpreted as follows: one of these three women's heights is 135, and two of these three women's heights is 139. In contrast, in the second group no woman was shorter than 140 cm, and there were four women (frequency column) whose height was between 140 and 145, namely: 143, 143, 143, and 144. The stem-and-leaf display shown in Table 2.5 was produced by PASW, which split the women who were in the 150–159 interval into two rows, because this interval is the most common in both groups. You will notice that the first of these two rows includes women whose height is up to 154, and the second includes women whose height is at least 155.

As you can see, stem-and-leaf displays are excellent tools for summarizing the samples. They are also a way to graphically show how the observations are distributed across different categories (in this case 140–149, 150–159, etc. cm). By looking at the length of the "leaves" attached to a stem, the reader can easily determine which of these stems is most common. It is obvious that in both groups, the most common interval is 150–159.

2.2 Graphing data

The ease with which graphs can be produced has had the unintended result that many papers or theses are submitted with uninformative graphs or graphs with useless information. The purpose of producing a graph is the same as that of producing a frequency distribution: displaying in a visual manner a summary of the data. Since graphs are produced to display data in an easily understandable manner, they must be simple, direct, and effective. If the reader needs to spend several minutes studying a graph to understand it, then the data analyst has done a poor job. For that reason, you should include different colors and a third dimension only when relevant, and you should try to keep the graph as simple as possible. In addition, you might want to sit down in front of your computer and play with it for a while. Whether you are using Excel, PASW, R, or SAS, you will not learn to use the program to its full potential in a few minutes or by reading a textbook. Moreover, it is not always obvious which type of graph will be most useful to you. Thus, you might want to try several before you decide on which one to show in a paper.

It does not hurt to remind ourselves that in a two-dimensional graph the **horizontal axis** is referred to as the **abscissa**, while the **vertical axis** is referred to as the **ordinate**. When we construct graphs in regression analysis, we will also refer to the former as the **X axis** and the latter as the **Y axis**.

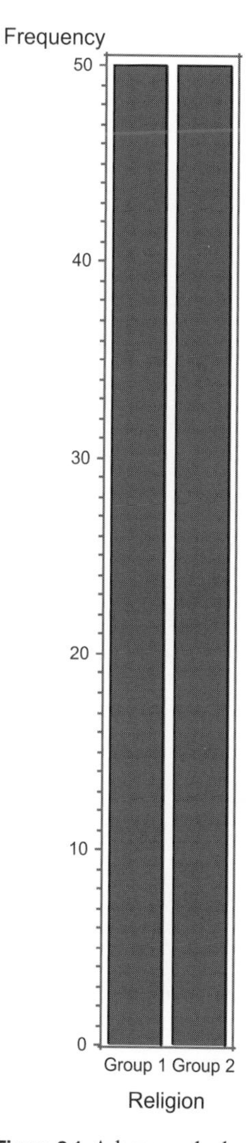

Figure 2.1 A bar graph showing the religious affiliation of a sample of 100 women.

2.2.1 Bar graphs and pie charts

Bar graphs are ideally suited for discontinuous numeric or qualitative variables. They show in a very clear manner the frequency at which specific observations occur. Just like we determined with a frequency distribution that some variates occurred more often than others, we can do the same by looking at the height of a bar. When qualitative variables are plotted, the different bars which represent the frequency of each observation should not be placed contiguously with each other. Figure 2.1 shows a (not-so-interesting) bar chart of religious affiliation in our sample of 100 women. You can clearly see that half of the women belong to one group and the other half to the other.

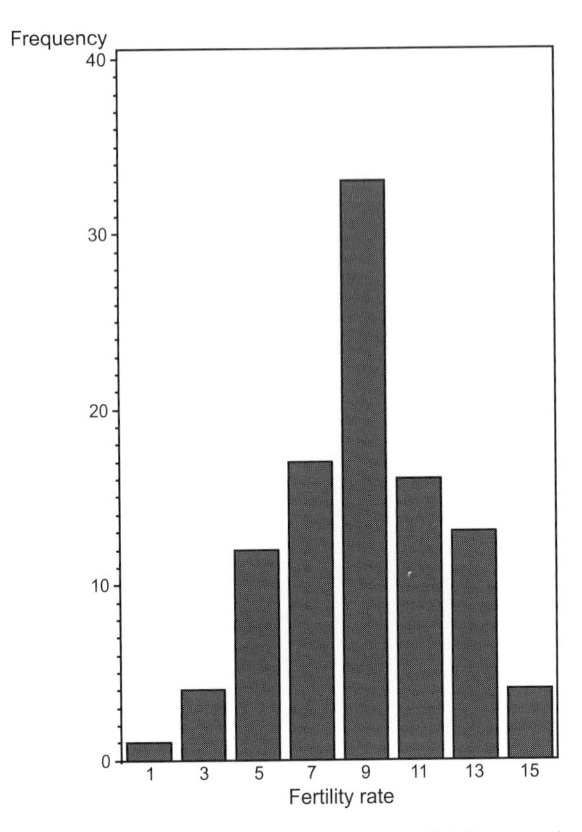

Figure 2.2 A bar graph showing the number of children produced by 100 women.

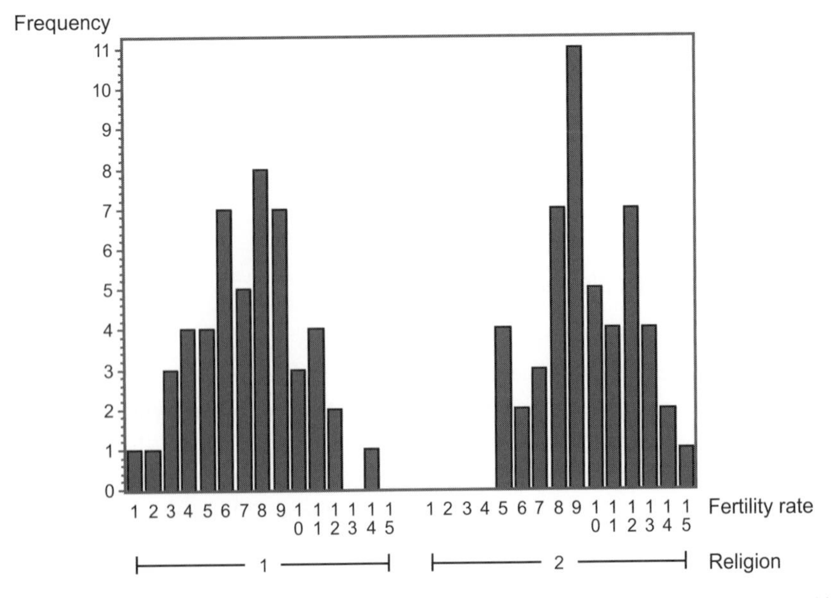

Figure 2.3 A bar graph showing the number of children produced by 100 women divided by their religious affiliation.

When a bar graph of discontinuous numeric variables is produced the bars representing the frequency of each outcome should not "touch" either, just as if we were working with a qualitative variable. Figures 2.2 and 2.3 show bar graphs of the number of children produced by all 100 women considered as a single group, and divided by religion (all of these were obtained by SAS). Figure 2.4 shows yet another option, namely, a horizontal bar chart, which in addition includes a frequency distribution of both samples. Although it is a bit "busy" this graph/chart combines a lot of useful information.

Pie charts are another popular choice when it comes to graphing discrete variables, either discontinuous numeric or qualitative. I used the data shown in Table 2.4 (women divided into height categories) to produce the pie chart shown in Figure 2.5.

2.2.2 Histograms

Histograms are rather similar to bar graphs, but are used for continuous numeric data. Since such data are in principle infinite (there are infinite numbers of possible heights between 170 and 171 cm), the bars of a histogram do "touch" each other. Most computer programs will allow you to choose how many bars to show and how narrow or broad those bars should be. In addition, both PASW and SAS allow you to superimpose a normal curve on your histogram, which will help you judge visually if your data depart dramatically from a normal distribution. What a normal distribution is, and how we test if a sample is normally distributed will be discussed below. Figure 2.6 is a histogram of weight of the women shown as a single sample, with a normal curve superimposed.

2.2.3 Polygons

These graphs are also applied to continuous numerical data, and are frequently preferred over histograms. Some computer packages refer to them as X*Y graphs, in which X represents the variable and Y the frequency of its various outcomes. Figures 2.7 and 2.8 show two polygons of height, for the women in our sample divided by religious affiliation.

2.2.4 Box plots

This is a very popular way to convey graphically basic information about the distribution of a sample. Although the construction of box plots requires an understanding of a few terms which are covered later, I will discuss it here since the necessary terms are very basic.

As a student, you probably took more than one standardized test and received your score and its percentile rank. If you were in the **25th percentile**, you did better than 25% of the other students who took the test. If you were in the **50th percentile**, you were exactly at the center of all the scores, or your score was at the **median** point of the distribution. If you were at the **75th percentile**, you did better than 75% of all the students. The difference between the 75th and the 25th percentile is known as the **interquartile range**. A box plot allows you to display the median and 25th and 75th

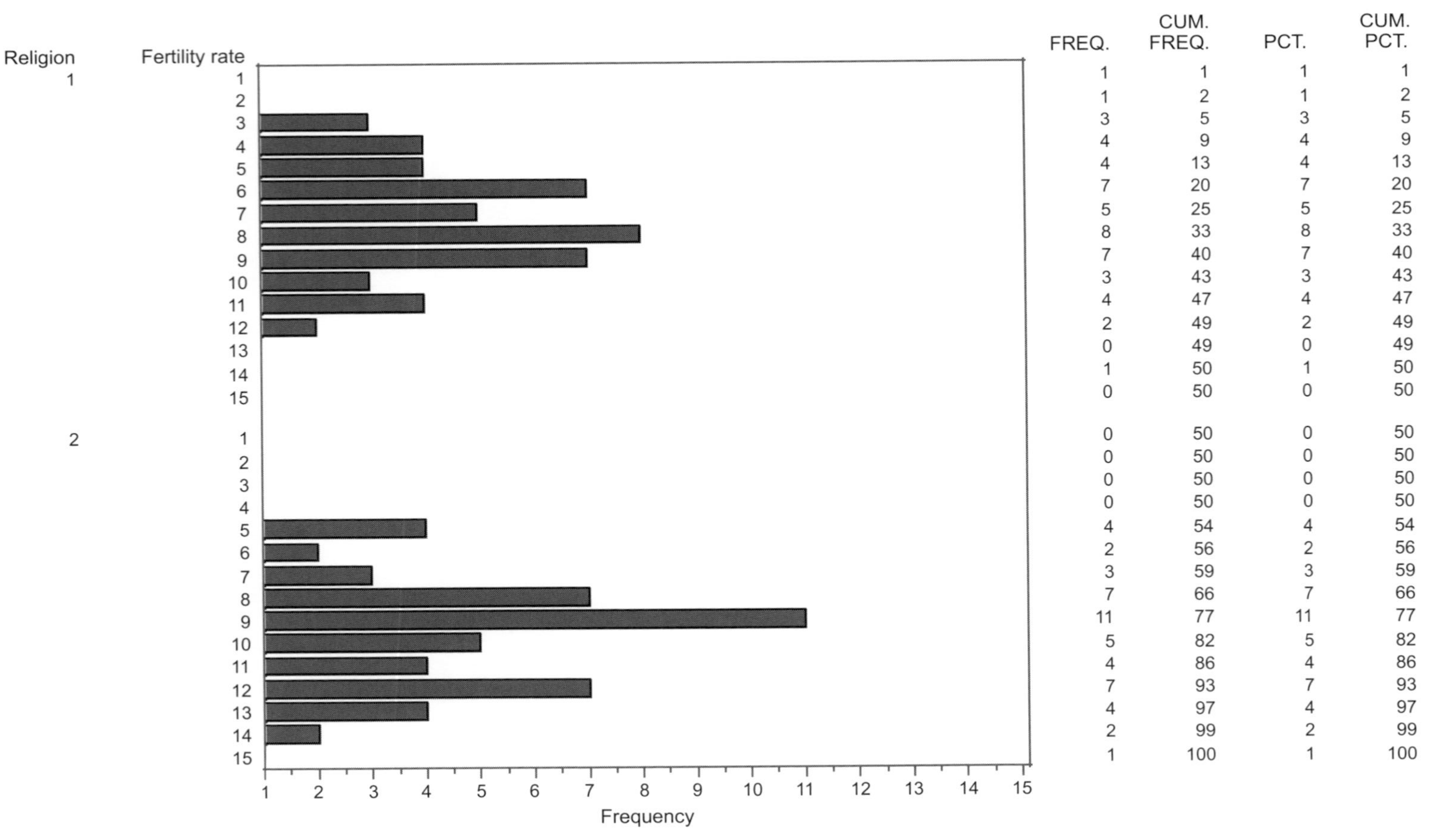

Religion	Fertility rate	FREQ.	CUM. FREQ.	PCT.	CUM. PCT.
1	1	1	1	1	1
	2	1	2	1	2
	3	3	5	3	5
	4	4	9	4	9
	5	4	13	4	13
	6	7	20	7	20
	7	5	25	5	25
	8	8	33	8	33
	9	7	40	7	40
	10	3	43	3	43
	11	4	47	4	47
	12	2	49	2	49
	13	0	49	0	49
	14	1	50	1	50
	15	0	50	0	50
2	1	0	50	0	50
	2	0	50	0	50
	3	0	50	0	50
	4	0	50	0	50
	5	4	54	4	54
	6	2	56	2	56
	7	3	59	3	59
	8	7	66	7	66
	9	11	77	11	77
	10	5	82	5	82
	11	4	86	4	86
	12	7	93	7	93
	13	4	97	4	97
	14	2	99	2	99
	15	1	100	1	100

Figure 2.4 A horizontal bar graph showing the number of children produced by 100 women divided by their religious affiliation.

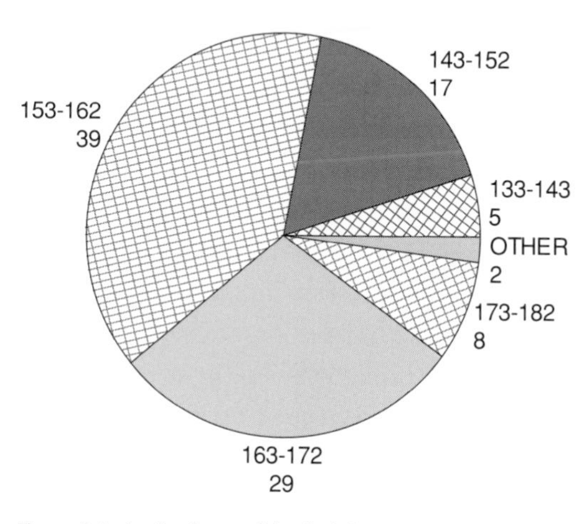

Figure 2.5 A pie chart of the height categories shown in Table 2.4. The category "other" includes two groups.

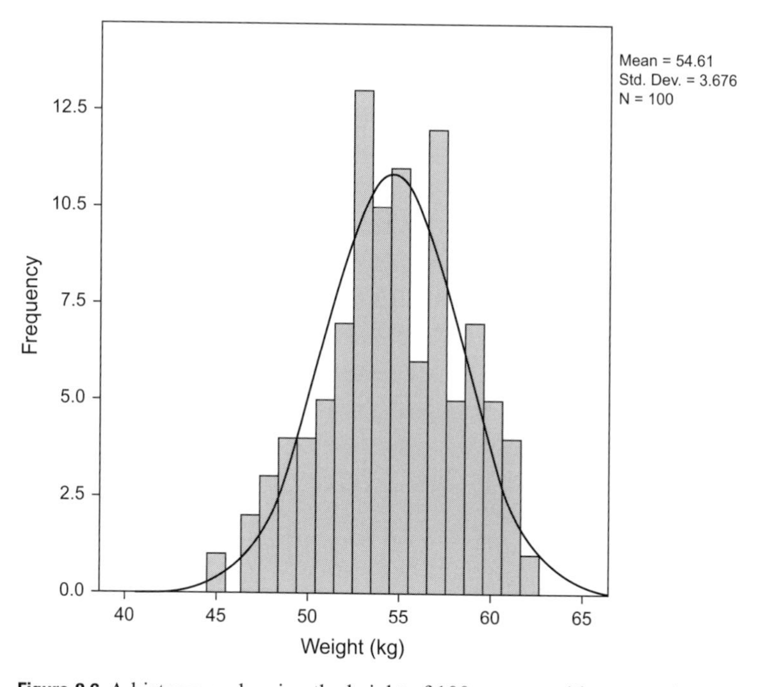

Figure 2.6 A histogram showing the height of 100 women with a normal curve superimposed.

percentiles of a sample. This constitutes the box. In addition to the box, a box plot also has vertical lines (the "whiskers") which show the location of the minimum and the maximum observation, as long as these observations are within 1.5 interquartile ranges from the top and the bottom. If there are more extreme observations, these are shown individually outside of the whiskers. Figure 2.9 shows a box plot of the number of children produced by women of both religious affiliations. The plot confirms what

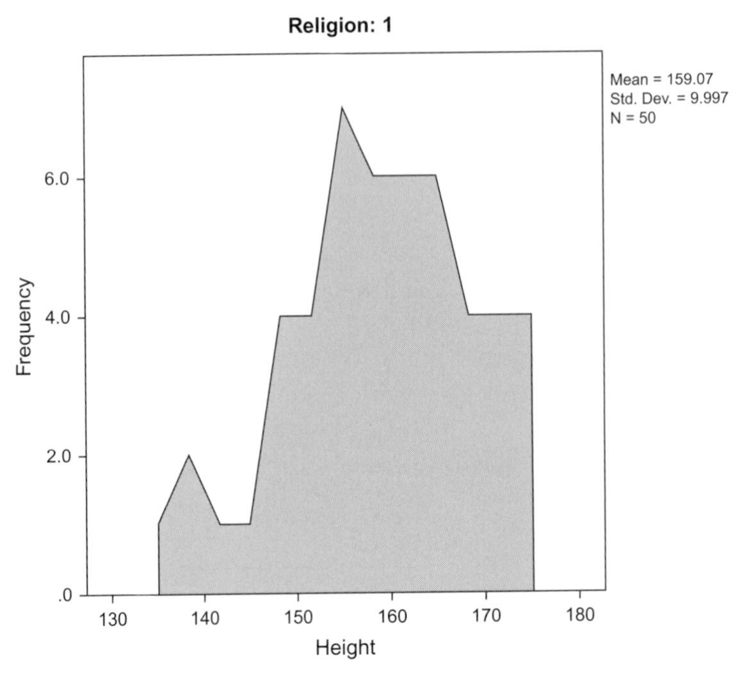

Figure 2.7 A polygon showing the distribution of height in 50 women who belong to group one.

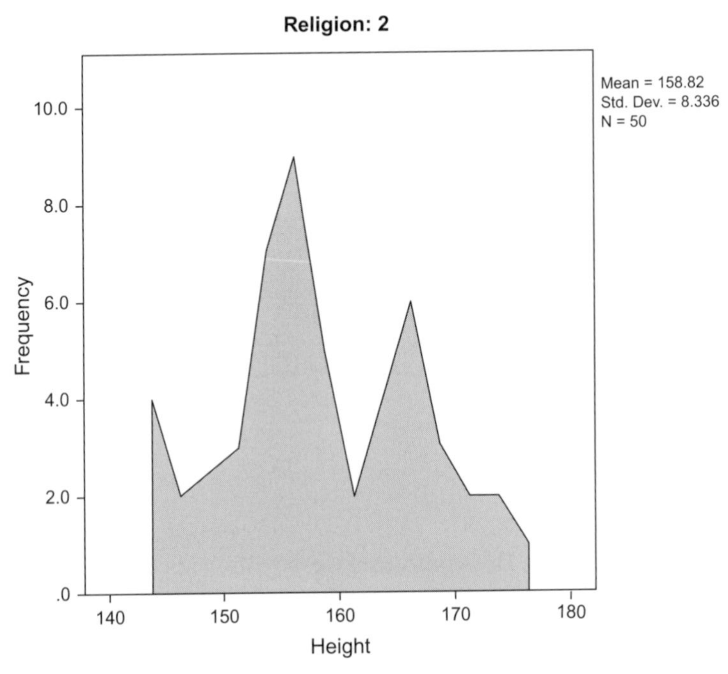

Figure 2.8 A polygon showing the distribution of height in 50 women who belong to group two.

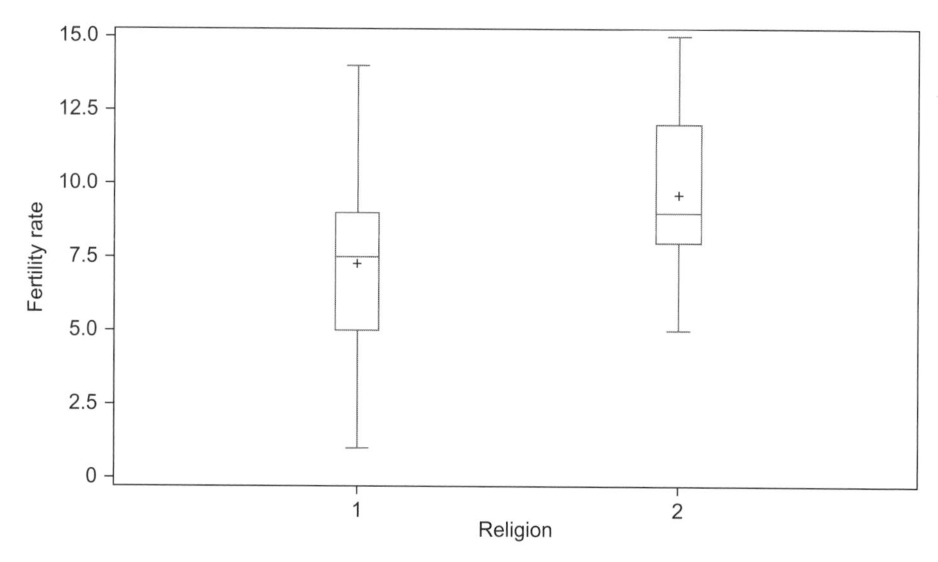

Figure 2.9 A box plot of the distribution of fertility in two groups of women divided by their religion.

we discussed previously: not only does group two have a higher median, but it also has higher minimum number of children produced.

2.3 Descriptive statistics. Measures of central tendency and dispersion

After constructing a frequency distribution and producing graphs, a researcher involved in quantitative analysis will usually compute sample descriptive statistics. Often, research papers do not include a frequency distribution or graph but only descriptive statistics. Descriptive statistics convey two basic aspects of a sample: **central tendency** and **dispersion**. The former describes the most representative or common or central observation of the sample, and the latter how the sample is distributed around the most common variate.

A word should be said about what calculation method may be easier in what situation. In a situation in which a researcher does not have access to a computer and needs to compute descriptive statistics of a large sample, he would probably choose to group the data into a frequency distribution because it is easier to compute the statistics in this manner. The problem with this approach is that if the data are grouped, the results will not be as precise because of rounding error (if two subjects' heights are 156.00023 and 155.999 cm you will probably assign them to the same category by rounding them to 156). With the use of computers, descriptive statistics are rarely computed by hand and, therefore, frequency distributions are rarely used for computation purposes. For that reason, I do not discuss how to compute descriptive statistics with frequency distributions. My assumption is that you will have a computer handy to be able to compute descriptive statistics with ungrouped raw data.

2.3.1 Measures of central tendency

This section discusses three central tendency statistics: the **mean**, the **median**, and the **mode**. The three are different kinds of "averages," used in different situations. Their general purpose is the same, namely, to find the single most representative score in the sample. Less commonly used measures of central tendency (such as the geometric and harmonic means) are covered by Sokal and Rohlf (1995).

2.3.1.1 The mean

This is the most commonly used statistic of central tendency for numeric continuous data. Indeed, when most people speak of "the average," they are actually referring to the mean. The mean considers every observation, and is therefore a statistic affected by extreme variates. Thus, for samples with extreme values, the median (see below) is more appropriate. A very important property of the sample mean is that it is a non-biased estimator of the **population mean** (μ) and is the central-tendency measure of choice in most cases.

The **sample mean** is commonly represented by a capital Y capped by a bar ($\bar{Y}$), and referred to as "Y-bar." In textbooks which favor the use of X instead of Y, $\bar{X}$ is preferred instead. The mean is computed by summing all the observations in a sample, and dividing the sum by the sample size. For example, if a student wants to obtain her mean test grade in a course in which 3 exams were administered (let us say her grades were 85, 90, and 100), the observations are summed ($85 + 90 + 100 = 275$), and divided by the sample size (3 in this case). Thus, the mean test score is $\frac{275}{3} = 91.7$. The only difference between the population and sample mean formulae is that in the former N (the population size), and in the latter n (the sample size) is used. The formula for computing the sample mean is given below:

Formula 2.1

Formula for the sample mean.

$$\bar{Y} = \frac{\sum Y}{n}$$

Let us compute the mean with the following data set:

5
6
8
7
10

100
8
9
6
10
10
5

First, add all the observations to obtain $\sum Y$: 5 + 6 + 8 + 7 + 10 + 100 + 8 + 9 + 6 + 10 + 10 + 5 = 184. Then, count the number of observations to obtain your sample size (n). In our example, $n = 12$. Finally, divide $\sum Y$ by n. Thus $\bar{Y} = \frac{184}{12} = 15.33$.

This example illustrates well the point made above: that the mean can be influenced by extreme values. In this data set all but one observation were below 11. However, because there was one observation whose value was 100, the mean is higher than 11, at 15.3.

Practice problem 2.1

The following is the age of 8 individuals attending a social service agency. Compute the mean age of the sample.

20
50
24
32
30
45
33
36

$\sum Y = 20 + 50 + 24 + 32 + 30 + 45 + 33 + 36 = 270$. Since $n = 8$, $\bar{Y} = 270/8 = 33.75$.

2.3.1.2 The median

The median is another measure of central tendency used with numerical variables. It is typically used with data which do not follow a normal distribution (see chapter three). For now, suffice it to say that the median is commonly used in data sets with extreme values. The median is the score that divides a distribution in half. Exactly one half of the scores are less than or equal to the median, and exactly one half are greater than or equal to it. This is the reason the median is not affected by extreme values. Readers are more likely to have seen reports of the median salary instead of the mean salary of a

Table 2.6 Type of paint in 50 archaeological pots.

Type	Code	Frequency	Percent	Cumulative frequency	Cumulative percent
Black	1	15	30.00	15	30.00
Plain	2	30	60.00	45	90.00
Red	3	5	10.00	50	100.00

sample. For example, in a university there is a large spread of salaries, from the salaries of the football coach to the salaries of undergraduate assistants. If a researcher were to obtain the mean salary at the university, she would be under the false impression that salaries are higher than they really are, because of the very high salaries received by some members of the university community. In this case, the analyst would be better off computing the median.

The median is easily computed by hand with a small sample size. With a large sample, you should use a computer program. We follow these steps to compute the median:

1. Order the variates from lowest to highest.
2. If sample size is odd, the median is the center-most value. For example, in the following data set: 6 8 5 101 9, order it from lowest to highest as follows: 5 6 **8** 9 101. The median in this case is 8, the center-most value in the sample.
3. If sample size is even, the median is the mean of the two center-most values (the $\frac{n}{2}$th and the $\frac{n}{2}$th + 1). For example, for the following data set:

1 3 90 2 10 26 44 73, order it from lowest to highest as follows: 1 2 3 **10 26** 44 73 90. The median is the mean of the two center-most values, thus: $\frac{(10+26)}{2} = 18$.

Practice problem 2.2

Compute the median with the following two sets of numbers:

30 90 80 150 5 0 160. The first task is to order the data: 0 5 30 **80** 90 150 160. The median is the center-most value: 80.

47 83 97 200 5 6. The first task is to order the data: 5 6 **47 83** 97 200. The median is the mean of the two center-most observations: $(47 + 83)/2 = 65$.

2.3.1.3 The mode

Archaeologists are frequently confronted with the task of classifying pottery according to types of decoration or color. Let us presume that a student has been hired by an archaeologist to classify the remains of 50 pots, and to write a short summary of his findings. The assistant assigns a code to each coloration, and constructs the frequency distribution shown in Table 2.6.

Which one is the average type? In this situation, the mode, or the most common value, should be reported. As a matter of fact, any time researchers deal with qualitative variables (gender, ethnic group, blood type, etc.), they should use the mode to report central tendency. In the example above, the modal ceramic type is the "Plain" one because it has the highest frequency (30). In some distributions, two categories have virtually equal frequencies. Such distributions are called bimodal, and have 2 modes (discussed in chapter three).

The mode can also be obtained for a frequency distribution of numerical variables. The mode is easily determined by examining the frequency distribution, and finding the category with the highest frequency. If you look at Table 2.1 (frequency distribution of number of children produced by women who belong to two religious groups) you will see that whereas the modal number of children produced by women in the first group is 8 (which has a frequency of 8), the modal number of children produced by women in the second group is 9 (which has a frequency of 11).

Practice problem 2.3

The following data are the height in 100 subjects, divided into intervals. What is the modal height interval?

Height	Frequency	Percent	Cumulative frequency	Cumulative percent
123–132	1	1.00	1	1.00
133–142	5	5.00	6	6.00
143–152	17	17.00	23	23.00
153–162	39	39.00	62	62.00
163–172	29	29.00	91	91.00
173–182	8	8.00	99	99.00
183–192	1	1.00	100	100.00

The mode in this example is the 153–162 category, which has the highest frequency (39).

2.3.2 Measures of variation

As mentioned previously, descriptive statistics summarize the sample by conveying the average observation, as well as the variation around it. A good measure of variability should provide an accurate picture of the spread of the distribution; in other words, it should give an indication of how accurately the average describes the distribution. If the variability is small, then the scores are close to the average. When variability is large, the scores are spread out, and are not close to the average.

2.3.2.1 The range

The range is the difference between the highest and lowest variates plus one. The one is added to acknowledge that continuous data (such as 156 cm) are really intervals that include higher and lower numbers such as 155.5 and 156.5. If we add the 0.5 of 155.5 and the 0.5 of 156.5 we obtain 1. It is important to keep in mind, however, that other statistics texts choose to define the range as the difference between the highest and lowest value without adding the one (Bohrnsted and Knoke 1988; Sokal and Rohlf 1995). Such difference of opinion is not of concern or much consequence. Indeed, SAS and PASW compute different values for the range of the same data set!

The range is used primarily when the researcher wants to convey how much difference there is between the lowest and highest score. I have found it useful when describing the spread of variation of live births in a sample of non-contracepting women: in my study the range was a whopping $17 - 0 + 1 = 18$ births (Madrigal 1989). However, as the range only takes into account 2 values, it does not express variation within the sample very well. By knowing that the range in my sample was 17, the reader still does not know how the fertility varies around the mean.

Practice problem 2.4

Compute the range for the number of cows owned by 6 families.

10
7
5
5
2
1

The range is $10 - 1 + 1 = 10$.

2.3.2.2 The population variance and standard deviation. The definitional formulae

Just like the mean, the population and sample variance and standard deviation have different symbols: The notation for the **sample variance** and the **standard deviation** is s^2 and s respectively. The notation for the **population variance** and **standard deviation** is σ^2 and σ respectively.

First, a few words about the variance and the standard deviation. The great advantage they have over the range is that they take into account all values in the population or in the sample instead of the two extremes. Therefore, they provide a much better idea of the amount of variation around the mean. The population variance is the mean squared deviation from the mean. In other words, the population variance computes the difference between each observation and the mean, squares it, and averages all the differences. Because the variance works with the squared distances from the mean, it is a large number relative to the original scores. To obtain a measure back in the original scale of measurement, the square root of the variance is computed. The standard deviation is the square root of the variance.

Of the two statistics, the standard deviation is more commonly used. Below are the formulae for the population variance and standard deviation. However the sum of squares (SS), also known as the corrected sum of squares (CSS) is first introduced. As its name implies, it is the sum of the squared differences between the observations and their mean.

Formula 2.2

Formula for the population sum of squares (SS).

$$SS = \sum (Y - \mu)^2$$

Formula 2.3

Formula for the population variance.

$$\sigma^2 = \frac{\sum (Y - \mu)^2}{N}$$

Formula 2.4

Formula for the population standard deviation.

$$\sigma = \sqrt{\sigma^2}$$

Let us assume that the following is a population, and that we wish to compute its variance and standard deviation.

Y
10
7
5
5
2
1

Let us follow these steps:

1. First, compute the mean. $\mu = 5$.
2. Second, obtain the difference between the mean and each observation.

Y	$(Y - \mu)$
10	$10 - 5 = 5$
7	$7 - 5 = 2$
5	$5 - 5 = 0$
5	$5 - 5 = 0$
2	$2 - 5 = -3$
1	$1 - 5 = -4$

3. Add the differences between the observations and the mean: $5 + 2 + 0 + 0 - 3 - 3 = 0$. If a population's or sample's mean is subtracted from each observation, and the differences are added, the result will be zero. This is why the differences must be squared.
4. Square the differences and add them up to obtain the sum of squares.

Y	$(Y - \mu)$	$(Y - \mu)^2$
10	$10 - 5 = 5$	25
7	$7 - 5 = 2$	4
5	$5 - 5 = 0$	0
5	$5 - 5 = 0$	0
2	$2 - 5 = 3$	9
1	$1 - 5 = 4$	16

Thus, $SS = \sum (Y - \mu)^2 = 54$.

5. Divide SS by N to obtain the population variance: $\sigma^2 = \frac{54}{6} = 9$.
6. To obtain the population standard deviation, compute the square root of the variance: $\sigma = \sqrt{9} = 3$.

Practice problem 2.5

Compute the variance and standard deviation of the following population.

Y	$(Y - \mu)$	$(Y - \mu)^2$
30	−4.67	21.8089
33	−1.67	2.7889
41	6.33	40.0689
43	8.33	69.3889
29	−5.67	32.1489
32	−2.67	7.1289

1. Compute the mean. $\mu = 34.67$.
2. Compute the sum of squares: $SS = \sum(Y - \mu)^2 = 173.3334$.
3. Divide the sum of squares by sample size (N = 6) to obtain the variance. $\sigma^2 = \frac{173.3334}{6} = 28.8889$.
4. Take the square root of the variance to obtain the standard deviation. $\sigma = \sqrt{\sigma^2} = \sigma = \sqrt{28.88} = 5.37$.

2.3.2.3 The sample variance and standard deviation. The definitional formulae

It is desirable for sample statistics to be non-biased estimators of population parameters. If sample statistics are biased they do not provide a good approximation of the true value of the population parameter. Whereas the sample mean is an unbiased estimator of the population mean, the sample standard deviation and variance are biased estimators of the population standard deviation and variance: they underestimate the value of the parameters. Therefore, the population formula cannot be applied to a sample when computing the sample variance and standard deviation.

Fortunately, the difference between the population and sample formulae is quite small: in the sample formula the sum of squares is divided by $n - 1$ instead of by n. The quantity $n - 1$ is known as the **degrees of freedom (df)**, and it will be used repeatedly later in the book.

Below are the formulae for the computation of the sample variance and standard deviation. The parameter notation is no longer used.

Formula 2.5

Formula for the sample variance.

$$s^2 = \frac{\sum(Y - \bar{Y})^2}{n - 1}$$

Formula 2.6

Formula for the sample standard deviation.

$$s = \sqrt{s^2}$$

To illustrate the difference between the population and sample formulae, let us compute the variance and standard deviation with the data sets we used previously to compute the population parameters.

Y
10
7
5
5
2
1

Let us follow these steps:

1. First, compute the mean: $\bar{Y} = 5$.
2. Second, obtain the difference between the mean and each observation.

Y	$(Y - \bar{Y})$
10	$10 - 5 = 5$
7	$7 - 5 = 2$
5	$5 - 5 = 0$
5	$5 - 5 = 0$
2	$2 - 5 = -3$
1	$1 - 5 = -4$

3. Square the differences and add them up to obtain the SS.

Y	$(Y - \bar{Y})$	$(Y - \bar{Y})^2$
10	$10 - 5 = 5$	25
7	$7 - 5 = 2$	4
5	$5 - 5 = 0$	0
5	$5 - 5 = 0$	0
2	$2 - 5 = -3$	9
1	$1 - 5 = -4$	16

Thus, $\sum(Y - \bar{Y})^2 = 54$.

4. Divide $\sum(Y - \bar{Y})^2$ by $n - 1$ to obtain the sample variance: $s^2 = 54/5 = 10.8$. Notice that the variance we computed with the population formula was lower, namely, 9.
5. To compute the population standard deviation, obtain the square root of the variance $s = \sqrt{10.8} = 3.28$. As you can see, the standard deviation obtained with the sample formula is also higher than that computed with the population formula ($\sigma = 3$). The population formula will produce values which are lower than those obtained by applying the sample formula.

Practice problem 2.6

Compute the variance and standard deviation of the following sample of ages of clients of a social service agency.

Y	$(Y - \bar{Y})$	$(Y - \bar{Y})^2$
30	−4.67	21.8089
33	−1.67	2.7889
41	6.33	40.0689
43	8.33	69.3889
29	−5.67	32.1489
32	−2.67	7.1289

1. Compute the mean. $\bar{Y} = 34.67$.
2. Compute the sum of squares: $\sum(Y - \bar{Y})^2 = 173.3334$.
3. Divide the sum of squares by sample size minus 1 ($n = 6$) to obtain the variance.

$$s^2 = \frac{173.3334}{5} = 34.67$$

4. Take the square root of the variance to obtain the standard deviation.

$$s = \sqrt{34.67} = 5.88$$

5. Compare these results with those obtained in practice problem 2.5.

2.3.2.4 The population and sample variance and standard deviation. The computational ("machine") formula

The population and sample formula for the variance and standard deviation presented in the previous sections are called the definitional formula because they define mathematically what these statistics are. With these formulae, we computed the difference between observations and their mean, squared them, added them, and divided their sum by N or $n-1$ to obtain the variance. Then we took the square root of the variance to compute the standard deviation. However, the definitional formulae are cumbersome and can easily accumulate rounding errors associated with adding squared differences. Therefore, it is better to use other formulae (called the computational or machine formulae) which are easier to work with, especially when working with large sample sizes and with a calculator. The standard deviation is computed as before, by taking the square root of the variance (see formulae 2.4 and 2.6).

Formula 2.7

Formula for the sum of squares (SS). "Machine" computation (use *n* if working with a sample, N if working with a population).

$$SS = \sum Y^2 - \frac{(\sum Y)^2}{n}$$

Formula 2.8

Formula for the population variance. "Machine" computation.

$$\sigma^2 = \frac{\sum Y^2 - \frac{(\sum Y)^2}{N}}{N}$$

Formula 2.9

Formula for the sample variance. "Machine" computation.

$$s^2 = \frac{\sum Y^2 - \frac{(\sum Y)^2}{n}}{n-1}$$

First, let us remind ourselves of what these statistics are: $\sum Y$ is the sum of the observations, $(\sum Y)^2$ is the sum of the observations squared, and $\sum Y^2$ is the sum of the squared observations. These operations must be performed in a specific order. First, square $\sum Y$ to obtain $(\sum Y)^2$. Then divide $(\sum Y)^2$ by sample (or population) size. Then, the quotient of the latter division should be subtracted from $\sum Y^2$. Finally, perform the last division by $n - 1$ if working with a sample or N if working with a population.

The difference between the sample and population variance is still that the sum of squares is divided in the former case by $n - 1$, and by N in the latter. To compute the standard deviation, the square root of the variance is taken (see formulae 2.4 and 2.6). The big difference between the machine and definitional formulae is the manner in which the sum of squares is computed. Let us use the data sets we used before, applying the sample formula. A column of the numbers squared is also included:

Y	Y^2
10	100
7	49
5	25
5	25
2	4
1	1

Let us follow these steps:

1. Compute $\sum Y$ and square it. $\sum Y = 30$, and $(\sum Y)^2 = 900$.
2. Square the numbers and add them up to obtain $\sum Y^2$. In this case, $\sum Y^2 = 204$.
3. The sample size is $n = 6$, and the degrees of freedom $n - 1 = 5$.
4. The variance is computed:

$$s^2 = \frac{204 - \frac{(30)^2}{6}}{6} = \frac{204 - \frac{900}{6}}{5} = \frac{204 - 150}{5} = \frac{54}{5} = 10.8$$

5. The standard deviation is the square root of the variance:

$$s = \sqrt{10.8} = 3.3$$

The value of the sum of squares, of the variance and the standard deviation, are all the same as computed with the definitional formula.

Practice problem 2.7

Compute the variance and standard deviation of the following sample of clients' ages attending a social service agency. Use the computational formula.

Y	Y^2
30	900
33	1089
41	1681
43	1849
29	841
32	1024

1. Compute $\sum Y$ and square it. $\sum Y = 208$, and $(\sum Y)^2 = 43{,}264$.
2. Square the numbers and add them up to obtain $\sum Y^2$. In this case, $\sum Y^2 = 7384$.
3. Keep in mind that $n = 6$, and $n - 1 = 5$.

$$s^2 = \frac{7384 - \frac{(208)^2}{6}}{5} = \frac{7384 - \frac{43{,}264}{6}}{5} = \frac{7384 - 7210.67}{5} = \frac{173.3334}{5} = 34.67$$

4. The standard deviation is the square root of the variance:

$$s = \sqrt{34.67} = 5.88$$

Practice problem 2.8

Compute the mean, median, mode, standard deviation, and range for the age, number of children produced (fertility), height, and weight for the data set called "women of two religions." Do this for the entire sample and for each group separately. Use the computer package of your choice. For this practice I used PASW. Here are the statistics for the entire sample. Please note that if there are several modes, PASW shows the smallest value of the modes (see footnote a).

Statistics					
		age	fertility	height	weight
N	Valid	100	100	100	100
Mean		34.26	8.47	158.95	54.58
Median		34.21	8.54	158.87	54.55
Mode		18[a]	1[a]	135[a]	45[a]
Std. Deviation		6.411	2.909	9.159	3.664
Range		29	13	41	16

[a] Multiple modes exist. The smallest value is shown.

Here are the statistics for the sample divided by religious membership.

	Religion		Statistic
fertility	1	Mean	7.3
		Median	7.48
		Variance	8.039
		Std. Deviation	2.835
		Range	13
	2	Mean	9.65
		Median	9.24
		Variance	6.255
		Std. Deviation	2.501
		Range	10

2.4 Chapter 2 key concepts

- Frequency distributions of qualitative variables.
- Frequency distributions of discontinuous numeric variables.
- Frequency distributions of continuous numeric variables.
- What is the purpose of constructing frequency distributions?
- What should you consider when building a graph?
- Review the notation of statistics and parameters discussed in this chapter.

- What is the difference between parametric and statistical formulae for the computation of the standard deviation?
- Why do we favor the "machine" formula instead of the definitional formula?
- Explain when the computation of the median and the mode are preferable to the computation of the mean.

2.5 Computer resources

1. I obtained my frequency distributions with SAS/ASSIST as follows: report writing → counts → list → one way. I find that if I have a data set which I like to split and de-split SAS is much more helpful than PASW.
2. I prefer to use PASW for obtaining descriptive statistics. You can obtain them following two different paths: Analyze → reports →, and Analyze → descriptive statistics → descriptives or Analyze → descriptive statistics → frequencies. You can also obtain simple descriptive statistics in SAS/ASSIST data analysis → elementary → summary stats. But if you want anything complex such as kurtosis and skewness you need to use proc univariate, which necessitates that you leave the menu-based format and write a code.
3. Graphs 2.1 and 2.7–2.9 were produced by PASW. All other graphs were produced by SAS.

2.6 Chapter 2 exercises

All data sets are available for download at www.cambridge.org/9780521147088.

1. Write down a research project of your own interest, describing the variables you wish to collect. How would you summarize/describe your data? Would you compute the mean or the median for numerical variables? What other descriptive statistics would you provide?
2. You conduct a survey on ethnic membership in a small community, and find the following numbers. Which measure of central tendency should you use and why? Compute it and show it next to a graph of your choice.

Ethnic group	f
Thai	30
Chinese	22
Laotian	11

3. Use the "infant birth weight" data set for the following:
   a. Compute the mean, median, mode, range, variance, and standard deviation for the variables birth weight, maternal weight, systolic blood pressure, and diastolic blood pressure. Do this for the entire sample.

b. Divide the sample into mothers who smoked and those who did not, and compute the same statistics for both groups.
c. Construct a frequency distribution of the variable smoked (*Y*/N) to determine the prevalence of this behavior in these mothers. Illustrate it with a graph.
d. Construct a histogram of systolic and diastolic blood pressure for the entire sample and then by smoking status.

4. Use the "school snacks" data set for the following:
   a. Compute the mean, median, mode, range, variance, and standard deviation for the variables height and weight. Do this for the entire sample.
   b. Divide the sample by gender and compute the same statistics for both groups.
   c. Divide the sample by activity level and compute the same statistics for both groups.
   d. Divide the sample by presence of snack machines and compute the same statistics for both groups.
   e. Construct a frequency distribution of the variable activity level and illustrate it with a graph.
   f. Construct a polygon for the entire sample and by activity level.

3 Probability and statistics

In this chapter we lay down the probability foundation for the rest of the book. All statistical tests rely on some aspects of probability, whether the test is working with qualitative data such as kinship titles or continuous data such as milliliters. The rules of probability are necessary for our understanding of statistical tests, but they are also necessary for our understanding of how scientific statements should be expressed, and more broadly, how science works. As I noted in chapter 1, science works with hypotheses which are submitted to tests and are either accepted or rejected but not proven. As we will see later, when we reject a hypothesis we reject it knowing full well that we could be committing an error and what the probability is that we are making an error. Therefore, the field of probability is essential to science-making because all predictions about the occurrences of events or explanations of events with hypotheses are qualified by probability statements. According to Fisher (1993), scientific inferences involving uncertainty are accompanied by the rigorous specification of the nature and extent of the uncertainty by which they are qualified. Since probability rules are fundamental to all scientific statements and to hypothesis testing, this chapter is extremely important to an understanding of statistical reasoning and hypothesis testing.

Many of you were exposed to the rules of probability for the first time in Elementary School. My students have told me that they learned the rules of probability in terms of blue and red marbles and that they forgot everything else about them except how pointless the marbles seemed at the time. Marbles may not be the best example to use, and I will try to focus on anthropological examples. However, the blue and red marbles examples remind us of the broad purpose behind learning probability in a statistics class. The purpose is for the student to understand that there are ways of predicting the likelihood that events or outcomes will occur.

Let us think about how the probability of outcomes or events may differ in different populations. Although there are variations across families, most English speakers use the term Aunt for father's sister and mother's sister. Therefore, if you are interviewing English speakers on what term they use for these relations, they will likely give you one answer: Aunt. In this case there is only one possible observation or event. However, many cultures do not give the same name to both sets of women and even see them as different types of relations. If you are in one of these cultures and you interview subjects on what term they use for these relations, the interview participants will give you two answers. In this second case there are two possible observations or events. The **sample space** is the set of all possible outcomes. In the first culture (culture A) the sample space only includes one event, whereas in the second culture (culture B) it includes two. The probability associated with any one event is defined by formula 3.1.

Formula 3.1

The probability of an event [Pr(E)].
The probability of an event is $0 \leq \Pr(E) \leq 1$

In culture A, where there is only one possible event the probability that these women will be called Aunt is 1. In culture B the probability associated with both possible events must be equal to 1. Therefore, if you do your interview in a household in which there are ten such women, and eight of them are father's sisters and two of them are mother's sisters, the probability of the first event is $\frac{8}{10} = 0.8$ and the probability of the second event is $\frac{2}{10} = 0.2$. If you add them up you obtain $0.8 + 0.2 = 1.0$.

If we are working with a population, we refer to the set of possible outcomes as the **probability distribution** of a variable, where the set of outcomes is unobservable because it describes the outcomes in a population (which is generally unobservable). What we hope to achieve is to obtain samples that accurately represent the true probability distribution of the variable in the population.

To summarize, the events we will consider here are observations of variables (qualitative, discontinuous numeric, and continuous numeric). We will discuss how to use data observed in a sample to compute the probability distribution of such events. For these observations to be recorded, they must first be sampled. We have already discussed what are desirable traits in a sample (representative and appropriate size), but we have not discussed how the sampling should occur. Therefore, this chapter is divided into the following major sections:

1. Random sampling and probability distributions.
2. The probability distribution of qualitative and discontinuous numeric variables.
3. The binomial distribution.
4. The Poisson distribution.
5. The Bayesian approach to probability and statistics.
6. The probability distribution associated with continuous variables (including *z* scores and percentile ranks).
7. The probability distribution of sample means and the Central Limit Theorem.

3.1 Random sampling and probability distributions

As I just mentioned, this chapter deals with the probability associated with events of different kinds, or the probability of obtaining a particular outcome in a particular population. But the first question we need to address is: how are we to sample from the population? We already noted that sampling should be **random**, that is, every individual should have an equal chance of being selected, which we express mathematically as $\frac{1}{N}$, which indicates that every individual in the entire population of size N has the same probability of being selected.

However, there is a feature of sampling which affects the probability of an individual being selected: sampling can be done with or without **replacement** of subjects. An excellent example of the difference between these two forms of sampling is the "shuffle" or "random" feature available in iPods. According to my older daughter, she has 2360 pieces of music in her iPod. When she puts the iPod in the "shuffle" feature, any one piece (let's call it #1) out of 2360 will be randomly selected and played. The probability of any one piece being selected is $\frac{1}{2360}$. I asked her if the second shuffle piece could be #1 again, or if #1 was out of the pool (she did not know). If piece #1 is out of the pool, then the iPod is doing **sampling without replacement**. This means that the probability of any one piece being selected as #2 is now $\frac{1}{2360-1}$ because there is one fewer piece in the population of pieces (#1 was removed). In the same manner, the probability of any one piece being selected as #3 is $\frac{1}{2360-2}$, and so on. Therefore, if a researcher is doing sampling without replacement, the probability associated with events does not remain stable but changes each time an observation is obtained because this last observation is not going to be put back into the population, whose size has just been diminished by one.

In contrast, if the iPod is doing **sampling with replacement**, then piece #1 is put back in the pool and it could very well be sampled again and again. Each time the probability that any piece of music will be selected remains stable, namely, $\frac{1}{2360}$. It is perfectly possible, although certainly not likely, that the same track will be played again and again, since each track has the same probability of occurring each time.

What should anthropologists do in the field when they are collecting data? Should they do sampling with or without replacement? Surely, we all hope to randomly obtain samples that are large and representative of the population. Thus, we may wish not to replace subjects so that we get enough different observations as to have an accurate representation in the sample of the population. For example, if an anthropologist is interested in interviewing female household heads to learn about the community's nutrition patterns, he would not want to interview the same female more than once (unless he wanted to check the reliability of his subjects' responses). Thus, the investigator would probably wish not to "replace" her, that is, not to give her a chance of being selected again. As long as the population from which the sample is obtained is large enough, sampling without replacement does not substantially change the initial probabilities associated with specific events in the population. Moreover, sampling without replacement may allow the investigator to obtain a sample which better represents the population.

3.2 The probability distribution of qualitative and discontinuous numeric variables

There are many kinds of qualitative and discontinuous numeric variables in anthropological research. They will both be referred to as discrete data since they can only take specific values (you can have blood type A, B, AB, or O, but nothing in between. You

Table 3.1 Ceramic types from Depot Creek shell mound (86u56). Test unit A. Data modified from White (1994).

Unit A	Frequency	Percent	Cumulative frequency	Cumulative percent
Check-stamped	185	58.17	185	58.17
Comp-stamped	8	2.52	193	60.69
Cord Mark	1	0.31	194	61.01
Grog-tempered	9	2.83	203	63.84
Indent-stamped	28	8.81	231	72.64
Sand-tempered	44	13.84	275	86.48
Simple-stamped	43	13.52	318	100.00

can own 0, 1, 2, or more goats, but not 0.75). The purpose of computing the probability of observing a specific outcome (say a type of hemoglobin) is to make inferences about the probability of sampling such outcomes from the population.

This topic is illustrated with an archaeological data set. In 1987 and 1988 White (1994) directed a number of excavations in various archaeological sites in the Apalachicola River Valley of Northwest Florida. The following data are the number of ceramic sherds by type found in test unit A at the Depot Creek shell mound (data modified from White 1994). Table 3.1 lists the number and percentages of ceramic types (White 1994).

The probability of sampling any of these outcomes is their frequency. Thus, the probability of sampling a check-stamped sherd is virtually 60%, whereas that of cord-marked is virtually 0. The computation of expected outcome probabilities is useful when planning a follow-up to a pilot study. If the researcher were to go back to the same site, and excavate in another area, and she found say, 60% cord, and 0.1% check-stamped sherds, she could conclude the following: (1) a very unlikely event has occurred, but the true distribution of ceramic types in this site is still the one predicted by the above table or (2) both test sites sampled a different population of ceramic types.

Practice problem 3.1

The following data show ceramic sherd types from Depot Creek shell mound (86u56) test unit B (White 1994). What is the probability p of selecting all seven outcomes? In your opinion, do test units A and B have similar enough frequencies as to indicate that they were sampled from the same population?

Although the two test units differ in their frequency of indent stamped pottery, they do appear to have a similar distribution for the other types. We will pursue later the issue of how exactly to test for this difference. However, at this point the reader realizes that knowledge of a probability distribution (test unit A) allows researchers to make statements about the likelihood of finding specific outcomes in another sample of the same population (test unit B).

Table 3.2 Ceramic types from Depot Creek shell mound (86u56). Test unit B. Data modified from White (1994).

Unit B	Frequency	Percent	Cumulative frequency	Cumulative percent
Check-stamped	88	40.74	88	40.74
Comp-stamped	8	3.70	96	44.44
Cord-stamped	4	1.85	100	46.30
Grog-tempered	15	6.94	115	53.24
Indent-stamped	46	21.30	161	74.54
Sand-tempered	40	18.52	201	93.06
Simple-stamped	15	6.94	216	100.00

Table 3.3 The number of horses owned by male heads of household in rural community A.

Number of horses	Frequency	Percent	Cumulative frequency	Cumulative percent
1	2	10.00	2	10.00
2	2	10.00	4	20.00
3	4	20.00	8	40.00
4	5	25.00	13	65.00
5	2	10.00	15	75.00
6	2	10.00	17	85.00
7	2	10.00	19	95.00
8	1	5.00	20	100.00

The reader recalls that discontinuous numerical variables have only fixed values. Thus, when computing probability distributions of discontinuous numerical data, the procedure is basically the same to that we just applied to qualitative data. The following example illustrates this. An anthropologist is performing a survey on the number of horses owned by 20 male heads of household in a small rural community. Table 3.3 shows the data. According to this table, the most frequent number of horses owned in the community is four, followed by three, with a probability of 0.25 (5/20) and 0.2 (4/20) respectively. The anthropologist could use this data to make predictions about the number of horses owned by men in the neighboring village. If he samples 20 men and finds that the frequency of men who own eight horses is 25%, then the anthropologist is likely to deduce that there is a different economic reality impacting the number of animals which men can own in both villages.

3.3 The binomial distribution

A frequently used distribution for qualitative variables in biological anthropology is the binomial distribution. Some of the fondest memories of many anthropology students are those of the computation of the Hardy–Weinberg equilibrium, which relies on the binomial distribution. The binomial expansion as most of you learned it applies to a situation in which there are two possible events (p and q or success and failure, for example).

Let us look at the binomial distribution with a familiar example. As you probably know, human populations are **polymorphic** for many genetic systems. That is, human populations have more than one allele for a specific locus. A well-studied genetic system is that for **hemoglobin** (Hb). There are many human populations which have other alleles besides "normal" hemoglobin (Hb A for adult hemoglobin). The most widely known abnormal hemoglobin is Hb S (sickle hemoglobin), although there are other hemoglobins such as Hb C and Hb E.

Let us assume that a population has the following allelic frequencies for hemoglobin: Hb A = 0.9, Hb S = 0.1 (thus, $\Pr(A) = 0.9$ and $\Pr(S) = 0.1$. We add $0.9 + 0.1 = 1.0$ to obtain the sample space in this example). When we talk about allelic frequencies we overlook that genes are carried by individuals, and visualize the population as a pool of genes, 90% of which consists of the A allele, and 10% of which consists of the S allele. Now we can ask: given these frequencies of alleles, what is the probability of sampling from this gene pool the following outcomes (genotypes or individuals): a homozygote Hb AA, a heterozygote or carrier Hb AS, and a homozygote Hb SS or sickle cell anemic person? We can visualize these probabilities as follows: an individual Hb AA has two A alleles, each of which has a frequency of 0.9. Therefore, the probability of sampling such an individual is $(0.9)(0.9) = (0.9)^2 = 0.81$. The same can be said of the probability of sampling a sickle-cell anemic individual with two Hb S alleles, whose probability is $(0.1)(0.1) = (0.1)^2 = 0.01$. Finally, the probability of sampling a heterozygous individual involves the sampling of an A and an S allele. But this can be done in either of two forms: A first and S second or S first and A second. Thus, the probability of sampling a heterozygote is $(0.9)(0.1) + (0.1)(0.9)$, or $2(0.9)(0.1) = 0.18$. Therefore, the probability of sampling the three genotypes is Hb AA = 0.81, Hb AS = 0.18, and Hb SS = 0.01. Note that the probabilities add up to 1. The reader has probably noticed that what we have done is the expansion of the binomial shown in formula 3.2.

Formula 3.2

Computation of probabilities associated with genotypic frequencies in a system with two alleles.

$$(p + q)^2 = p^2 + 2pq + q^2$$

Why do biological anthropologists use the binomial expansion? In studies of gene frequencies, the expected frequencies generated by the binomial expansions are those we would expect to find if the population is in Hardy–Weinberg equilibrium, in other words, if the population is not evolving and mates randomly. If we find a statistically significant difference between the observed and expected frequencies, we can conclude that the population is either experiencing evolutionary change or is not mating at random.

Practice problem 3.2

In an attempt to study the genetic structure of the present-day Mexican population, Lisker *et al.* (1996) compute admixture estimates relying on gene frequencies thought to be representative of the parental populations. The authors present a table of gene frequencies of these ancestral populations, namely, American Indian, African, and European. For the American Indian population, they report the following gene frequencies for the MN system: M = 0.698 and N = 0.302. Compute the expected genotype frequencies.

If p is the frequency of the M allele, and q is the frequency of the N allele, then the expected genotypic frequencies are: $(p + q)^2 = p^2 + 2pq + q^2 = (0.698)^2 + 2(0.698)(0.302) + (0.302)^2 = 0.49 + .0.42 + 0.09 = 1$.

Notice how these expected frequencies could be used. If a biological anthropologist thought that these frequencies reflected the true frequencies of Mesoamerican populations before contact with Europeans and Africans (an assumption for this investigation), and if he wanted to test if an American Indian group had no admixture, then he would compare these expected frequencies with those observed in the population. If the researcher finds that the observed frequencies do not differ significantly from those expected, he can conclude that this particular group does not differ from the ancestral population in its MN frequencies. If the observed and expected frequencies differ, the researcher would conclude that the population has experienced evolutionary change making it different from the ancestral population. Such evolutionary change could have resulted from gene flow, genetic drift, etc. In reality, these studies rely on many, not just one, genetic systems.

3.4 The Poisson distribution

If you are into pescetarianism or love French fish recipes, please do not be disappointed by the true subject of this section: a theoretical distribution of discrete events. You recall that the binomial distribution is usually applied in situations with two possible outcomes (success or failure, p or q). Although the binomial can be expanded to deal with more than two outcomes (three hemoglobin types, for example), it quickly becomes very difficult to handle with multiple possible outcomes. The Poisson distribution deals with discrete events recorded over time or over geography (the sampling unit). But the Poisson distribution allows for many more possible outcomes, and for a frequency of zero in many of the sampling units. For this reason, the Poisson distribution can be seen as a distribution of rare discrete events. Indeed, its first description by Simeón-Denis Poisson deals with the frequency of mule kicks (a fortunately rare event) in the French army (Griffith and Haining 2006).

When we discussed the probability distribution of discrete variables we used the data obtained in one sample (test unit A) to make predictions about what we would find in another one (test unit B). When we discussed the binomial distribution we computed the genotypic frequencies expected if a population is not evolving to compare them with the observed frequencies in the population. In other words, we have used observed frequency distributions and compared them with theoretical probability functions (generated by the binomial or by observations in a test unit) to see if there is a difference between them. If the observed genotypic frequencies do not differ from those derived by the theoretical expectation, then we accept the proposition that the population is not evolving. If the frequency of ceramic sherd types does not differ between both test units then we accept the proposition that they reflect the same population of ceramic types. Our purpose in generating a Poisson distribution is to determine if the frequency of rare events we witness could have been generated by a theoretical Poisson distribution. We can describe the Poisson distribution and its assumptions as follows:

- It describes the distribution of the variable Y, which occurs as counts. Some anthropological examples are the number of projectile points per square meter, the number of influenza deaths per month, the number of prey encountered in a day by a hunter, etc.
- The variable Y is measured over an appropriate unit of time or geography. Therefore, while it makes sense to count the number of projectile points over a square meter, it does not make any sense to count it over a square centimeter. While it makes sense to record the number of influenza deaths over a month, it does not make sense to do so over a minute.
- The mean number of events must be low in comparison with the maximum number of events per sampling unit. Remember that we are working with rare events such as mule kicks. The reason the mean number of events is low is that there are several sampling units in which the frequency of the event is 0 or 1 (many months in which no one was kicked by a mule).
- The occurrence of events within a sample unit is independent of the occurrence of prior events in the same unit. In other words, the events are independent. You can see that this assumption of the Poisson distribution is most likely to be violated in the case of an influenza epidemic, in which new cases are tied to prior cases because there was contagion between the patients. But this is precisely why the Poisson distribution is so useful: it allows us to test if rare events occur independently, as assumed by the theoretical distribution. There are two ways in which the assumption of independence can be violated, one of which we already mentioned: (1) The cases occur in "clumps" so that the observed distribution suffers from "contagion" or "**clumping**." (2) The occurrence of one case precludes the occurrence of another one in another sampling unit. The observed distribution in this situation would be referred to as a **repulsed distribution**, and it could result from (for example) a prey animal of a certain species traveling alone. If an animal is encountered over a unit time of hunting then we are not likely to encounter another animal in the next unit because the occurrence of the event "repulses" another occurrence.

- In a Poisson distribution $\mu = \sigma^2$. If the observed distribution departs sharply from this expectation, then it is unlikely to be following a Poisson distribution. An easy way to check if this assumption holds in an observed distribution is to compute the **coefficient of dispersion** as follows: $CD = s^2/\bar{Y}$. If an observed distribution of rare events does follow the theoretical Poisson distribution, its CD should be close to 1. However, to test with certainty if an observed distribution follows the Poisson distribution, we need to wait until we cover tests for goodness-of-fit.

To summarize, we use the Poisson distribution to compute expected frequencies of rare events. These frequencies are expected if the events are independent of each other. Therefore, the Poisson distribution is a great way of testing if our expectation of independence in the rare event of interest is true or not.

The computation of the expected Poisson frequencies necessitates the use of the exponential and factorial functions in your calculator. It seems that each calculator has a different exponent key, so I will not attempt to explain how to use it. However, I do want to remind you that the factorial of 2 is 2! = 2*1 = 2, the factorial of 3 is 3! = 3*2*1 = 6, the factorial of 4! = 4*3*2*1 = 24, and so forth.

While some authors prefer to compute the **expected raw frequency** (without considering sample size), others prefer to compute the **absolute expected frequency** (the expected frequency considering the sample size). Fortunately, we do not have to do all of this by hand, as PASW allows us to compute both frequencies very easily. This is discussed at the end of the chapter under "computer resources." We present the formulae for the computation of both below, with a full example in practice problem 3.3.

Formulae 3.3–3.4

Computation of the relative and absolute expected frequencies in a Poisson distribution. Only seven possible events are shown.

Counts observed (Y)	Observed frequencies (f)	Formulae for computation of relative expected frequencies ($\hat{f}_r$)	Formulae for computation of absolute expected frequencies ($\hat{f}$)
0	f_o	$\hat{f}_{r0} = \frac{e^{-\bar{Y}}(\bar{Y})^0}{0!}$	$\hat{f}_0 = \frac{n}{e^{\bar{Y}}}$
1	f_1	$\hat{f}_{r1} = \frac{e^{-\bar{Y}}(\bar{Y})^1}{1!}$	$\hat{f}_1 = \hat{f}_0\,\bar{Y}$
2	f_2	$\hat{f}_{r2} = \frac{e^{-\bar{Y}}(\bar{Y})^2}{2!}$	$\hat{f}_2 = \hat{f}_1\frac{\bar{Y}}{2}$
3	f_3	$\hat{f}_{r3} = \frac{e^{-\bar{Y}}(\bar{Y})^3}{3!}$	$\hat{f}_3 = \hat{f}_2\frac{\bar{Y}}{3}$
4	f_4	$\hat{f}_{r4} = \frac{e^{-\bar{Y}}(\bar{Y})^4}{4!}$	$\hat{f}_4 = \hat{f}_3\frac{\bar{Y}}{4}$
5	f_5	$\hat{f}_{r5} = \frac{e^{-\bar{Y}}(\bar{Y})^5}{5!}$	$\hat{f}_5 = \hat{f}_4\frac{\bar{Y}}{5}$
6	f_6	$\hat{f}_{r6} = \frac{e^{-\bar{Y}}(\bar{Y})^6}{6!}$	$\hat{f}_6 = \hat{f}_5\frac{\bar{Y}}{6}$
7	f_7	$\hat{f}_{r7} = \frac{e^{-\bar{Y}}(\bar{Y})^7}{7!}$	$\hat{f}_7 = \hat{f}_6\frac{\bar{Y}}{7}$

Total number of occurrences $= \sum fY = \sum y_i * f_i + \cdots + y_k * f_k$ for all k outcomes.
Total sample size $= n = \sum f$.
Estimate of μ is $\bar{Y} = \frac{\sum fY}{n}$.
Estimate of σ^2 is $s^2 = \frac{\sum Y^2 - \frac{(\sum Y)^2}{n}}{n-1}$.

Practice problem 3.3

An unfortunate graduate student is being asked to survey a corridor of land which is being proposed as an area for the construction of a new road in order to determine if there are archaeological remains. Because previous work indicates that in this specific region projectile points are most likely to be preserved, the student is going to count projectile points. He divides the area into 12 one-square meter sampling units and records the number of projectile points he finds. Notice that the Y (our variable of interest) is the number of points recorded, while the observed frequency is the number of square meters where such numbers were observed. Thus, the student found zero projectile points in two sampling units, one point in two units, two in three units, etc. Compute the relative expected and absolute expected frequencies.

Total number of occurrences $= \sum fY = 0{*}2 + 1{*}2 + \ldots 6{*}1 + 7{*}1 = 33$.
Total sample size $= n = \sum f = 2 + 2 + \ldots 1 + 1 = 12$.
Estimate of μ is $\bar{Y} = \frac{33}{12} = 2.75$.
Estimate of σ^2 is $s^2 = \frac{149 - \frac{33^2}{12}}{11} = 5.29$. These numbers were obtained as follows:

$$\sum fY = 0 + 0 + 1 + 1 + 2 + 2 + 2 + 3 + 4 + 5 + 6 + 7 = 33.$$
$$\sum Y^2 = 0^2 + 0^2 + 1^2 + \ldots + 5^2 + 6^2 + 7^2 = 149.$$

Y	Obs. freq. (f)	Relative expected frequencies	Absolute expected frequencies ($\hat{f}$)
0	2	$\hat{f}_{r0} = \frac{e^{-2.75}(2.75)^0}{0!} = 0.0639$	$\hat{f}_0 = \frac{\sum fY}{e^{2.75}} = \frac{33}{15.64} = 2.1096$
1	2	$\hat{f}_{r1} = \frac{e^{-2.75}(2.75)^1}{1!} = 0.1758$	$\hat{f}_1 = \hat{f}_0\,\bar{Y} = 2.1096{*}2.75 = 5.8015$
2	3	$\hat{f}_{r2} = \frac{e^{-2.75}\cdot(2.75)^2}{2!} = 0.2417$	$\hat{f}_2 = \hat{f}_1 \frac{\bar{Y}}{2} = 5.8015{*}\frac{2.75}{2} = 7.9770$
3	1	$\hat{f}_{r3} = \frac{e^{-2.75}(2.75)^3}{3!} = 0.2215$	$\hat{f}_3 = \hat{f}_2 \frac{\bar{Y}}{3} = 7.9770{*}\frac{2.75}{3} = 7.3122$
4	1	$\hat{f}_{r4} = \frac{e^{-2.75}(2.75)^4}{4!} = 0.1523$	$\hat{f}_4 = \hat{f}_3 \frac{\bar{Y}}{4} = 7.3122{*}\frac{2.75}{4} = 5.0272$
5	1	$\hat{f}_{r5} = \frac{e^{-2.75}(2.75)^5}{5!} = 0.0837$	$\hat{f}_5 = \hat{f}_4 \frac{\bar{Y}}{5} = 5.0272{*}\frac{2.75}{5} = 2.7649$
6	1	$\hat{f}_{r6} = \frac{e^{-2.75}(2.75)^6}{6!} = 0.0384$	$\hat{f}_6 = \hat{f}_5 \frac{\bar{Y}}{5} = 2.7649{*}\frac{2.75}{6} = 1.2673$
7	1	$\hat{f}_{r7} = \frac{e^{-2.75}(2.75)^7}{7!} = 0.0151$	$\hat{f}_7 = \hat{f}_6 \frac{\bar{Y}}{7} = 1.2673{*}\frac{2.75}{7} = 0.4979$

Here I calculate a few of the relative expected frequencies:

$$\hat{f}_{r1} = \frac{e^{-2.75}(2.75)^1}{1!} = 0.0639{*}2.75 = 0.1758$$

$$\hat{f}_{r2} = \frac{e^{-2.75}\cdot(2.75)^2}{2!} = \frac{0.0639{*}7.5625}{2} = 0.2417$$

$$\hat{f}_{r3} = \frac{e^{-2.75}(2.75)^3}{3!} = \frac{0.0639{*}20.7968}{6} = 0.22158$$

Sum of absolute expected frequencies $= 2.1096 + 5.8015 + \ldots + 1.2673 + 0.4979 = 32.7 \approx 33$.

The astute reader surely noticed that s/he does not have to do all calculations shown above. Once you have computed the relative expected frequency, you can easily obtain the absolute expected frequency by multiplying the former by $\sum fY$. Thus for $f_0 = 0.06393{*}33 = 2.10969$, $f_1 = 0.1758{*}33 = 5.8014$, and so forth.

$\mathrm{CD} = \frac{5.29}{2.75} = 1.92$. The coefficient in this example is not particularly close to 1. It is likely that the distribution of projectile points does not follow a Poisson distribution.

Applying the Poisson distribution in anthropology

The Poisson distribution may be of use to anthropologists in different areas of research, as long as the research project deals with rare occurrences of events. An obvious topic is that of the geographical distribution of archaeological remains. If we find that the distribution of certain artifacts departs from the Poisson expectation but demonstrates a "clumping" distribution, then the distribution likely indicates that these artifacts were either used or manufactured in designated areas instead of throughout the entire site. Readers interested in the application of the Poisson distribution throughout geographical spaces should see the paper by Griffith and Haining aptly entitled "Beyond mule kicks: the Poisson distribution in geographical analysis" (Griffith and Haining 2006).

The Poisson distribution has been applied to test a central assumption of the optimal foraging theory when applied to human hunting: that prey items are encountered sequentially as a Poisson process (Zeleznik and Bennett 1991). Zeleznik and Bennett (1991) studied encounters with prey of Bari hunters in Venezuela. The authors compared the observed frequencies of encounters with Poisson expected frequencies, and concluded that the observed and expected frequencies did not differ significantly. Zeleznik and Bennett (1991) concluded that since the Poisson distribution assumes that the events are rare and independent, and the distribution of prey encounters does not differ from it, the Bari hunters are unable to manipulate the environment in order to "clump" the encounters with prey, which would be beneficial for the hunters (more animals are encountered in one hunting party).

Many students of epidemiology have used the Poisson distribution to determine if new cases of diseases are independent or if they "clump" because the disease is contagious. The Poisson distribution can be applied to a temporal study of disease distribution (do we see the same number of new cases each week in a month, or each month in a year?) or to a geographical study of disease distribution (do we see the same number of new cases in all areas of the city or do new cases appear preferentially in certain areas?). The Poisson distribution can even be useful for calculating expected frequencies of new cases in the absence of sufficient data on the history of the disease in a particular region (Sartorius *et al.* 2006).

3.5 Bayes' theorem

We discussed the binomial distribution, in which events appear as success or failure (p or q), and we estimate the probability associated with these different outcomes. We also discussed the Poisson distribution, in which events are assumed to be rare and independent of each other, and which allows us to determine if these assumptions are met in real data. Bayes' theorem also deals with data that are counts such as the number of people with the HB SS genotype, or the number of projectile points found in one square meter. What is highly desirable about a Bayesian approach to probability is that it allows us to incorporate prior knowledge about an event to have a better estimate of its occurrence. In a Bayesian approach to probability, there is the **prior probability**, which is the best estimate a researcher has about the frequency of an event. For example, you may guess that the *salsa* club which you just joined has half local and half foreign students in its membership. This is your prior probability. Once you visit the club, you learn that your university has an exchange program with Caribbean countries, which makes the university rather affordable to students of these countries. As a result of the exchange, your *salsa* club actually has 60% foreign and 40% local students. These latter numbers are the **posterior probability**, which is the probability of the events computed after the researcher has gathered additional information about the events.

In most books Bayes' theorem is explained using an epidemiological example, and I am not going to go against this tradition. However, in the practice problem I will use a non-clinical data set to illustrate the use of Bayes' theorem in other research settings.

Bayes' theorem has been exceedingly helpful in evaluating the proportion of correct (positive) diagnosis in individuals who do have a disease, or the **positive predictive value**. In other words, if you are concerned about suffering from a disease, you might find out how prevalent the disease is in the general population, or in people of your gender and age. That should give you an idea of what your chances are of getting the disease. Let us say that you go a step ahead and take the test to diagnose the disease and the test comes out positive. Bayes' theorem is going to allow you to determine the proportion of people who do have the disease and whose test does come out positive. We need to remember that tests for the detection of disease are not always desirable because they sometimes fail to detect the disease (**false negative**) or because they erroneously indicate the presence of the disease in a healthy person (**false positive**). Indeed, since 2009 there has been sizable controversy about the frequency with which women

40–49 years of age should have mammograms because many tests yield false positives causing undue mental anguish on these patients (Calonge *et al.* 2009; Greif 2010).

For us to be able to work out an example, we need to define a few terms. The **sensitivity** of a test is the probability of a positive test in a person who actually has the disease (a man who has prostate cancer is given a diagnostic test for the disease and he is correctly diagnosed as having it). The **specificity** of a test is the probability of a negative result in an individual who does not have the disease (a man who does not have prostate cancer is given a diagnostic test for the disease and he is correctly diagnosed as not having it). The **prevalence** of the disease is defined as the proportion of subjects who have the disease in a population. The question of who is "the population" is not always clear. Is it all men over age 20? Is it men in the age category of the man in question? How large or small should that category be?

The notation for probability statements is as follows: the probability of an event A is P[A]; the probability of an event B is P[B]; the **conditional probability** of A given B is P[A|B]. With this information we refer to possible outcomes as follows:

Probability of being diseased: P[D]. This is usually derived from population measures such as the proportion of men of a certain age that have the disease, better known as **the prevalence of the disease**.

Probability of not being diseased: P[ND]
Probability of positive tests among diseased individuals (**sensitivity**): $P[T^+|D]$
Probability of negative tests among non-diseased individuals (**specificity**): $P[T^-|ND]$
Probability of negative results among diseased individuals (**false negatives**): $P[T^-|D]$
Probability of positive results among non-diseased individuals (**false positives**): $P[T^+|ND]$

Bayes' theorem allows us to ask a much more relevant question than "What is the prevalence of the disease?" or "What proportion of all tests is positive?" Instead, we can ask: "Out of those tests which are positive, how many of them truly correspond to someone who is diseased?" If we keep in mind the controversy surrounding diagnostic tests for prostate, breast, and cervical cancer, you can see that this is a much more useful question (to the patient). We write the theorem as follows:

Formula 3.5

Computing the positive predictive value with Bayes' theorem.
The probability of being diseased, given a positive result is:

$$P[D|T^+] = \frac{P[D]^* \, P[T^+|D]}{P[D]^* \, P[T^+|D] + P[ND]^* \, P[T^+|ND]}$$

Once you have established the prevalence of the disease (P[D]), the sensitivity of the test $P[T^+|D]$ (these two terms appear in the numerator and the denominator), the complement of the prevalence of the disease (P[ND] = 1 − P[D]) and the proportion of false positives of the test (P[T$^+$ |ND]) you can compute the positive predictive value without much problem. Let's practice with a fictitious example:

Probability of being diseased or prevalence of disease: P[D] = 0.00007
Probability of not being diseased: P[ND] = 1 − 0.00007 = 0.99993
Proportion of positive tests among diseased individuals (sensitivity): P[T$^+$|D] = 0.9
Proportion of negative tests among non-diseased individuals (specificity): P[T$^-$|ND] = 0.85
Proportion of negative results among diseased individuals (false negatives): P[T$^-$|D] = 0.16
Proportion of positive results among non-diseased individuals (false positives): P[T$^+$|ND] = 0.16

Let us compute the probability of a person being diseased, given a positive result:

$$P\left[D|T^+\right] = \frac{0.00007*0.9}{0.00007*0.9 + 0.99993*0.16}$$
$$= \frac{0.000063}{0.000063 + 0.1599888} = \frac{0.000063}{0.1600518} = 0.0003936.$$

Let us examine what benefit we get from doing this extra work (besides the sheer fun of doing it). While the prevalence of the disease in individuals of the same age as our patient is 0.00007, the probability that a diseased patient will be detected by the test is much higher: 0.0003936. Therefore, the patient and his medical practitioner can be quite sure that the test really detected the disease. This example is perhaps a bit unrealistic because the specificity and the sensitivity are unrealistically high. The higher these numbers, the better the test is for detecting true cases and for not reporting false cases.

Practice problem 3.4

Computing the positive predictive value with Bayes' theorem.
Enamel hypoplasias are a well-known marker of childhood stress (the enamel deposition is disrupted) in skeletal populations. An anthropology student is interested in determining to what extent the presence of enamel hypoplasias in children's teeth is noted by a general dentist check-up as opposed to one done by the student himself. Our student wants to know to what extent this important tool to understand stress in skeletal populations is also of use in current pediatric dental practice. His population is all children of age 10 in the country. He needs the following information:

Probability of being diseased or prevalence of enamel hypoplasias in all children of age 10 (taken from the government health statistics): P[D] = 0.10
Probability of not being diseased: P[ND] = 1 − 0.10 = 0.90
Proportion of positive tests among diseased individuals, or determining that a child has hypoplasias when in fact s/he does (sensitivity): P[T$^+$|D] = 0.95

Proportion of negative tests among non-diseased individuals, or correctly determining that a child does not exhibit hypoplasias (specificity): $P[T^-|ND] = 0.80$

Proportion of negative results among diseased individuals (false negatives): $P[T^-|D] = 0.10$

Proportion of positive results among non-diseased individuals (false positives): $P[T^+|ND] = 0.14$

The probability of having enamel hypoplasias, given a positive test result is:

$$P[D|T^+] = \frac{0.1*0.95}{0.1*0.95 + 0.9*0.14} = \frac{0.095}{.221} = 0.42986.$$

Bayesian probability in anthropology

There are two areas in anthropology in which the use of the Bayesian approach to probability has grown dramatically in the recent past. The first is forensic anthropology, specifically when providing a probability statement associated with the identification of some skeletal remains. For example, we may know that the probability of finding a male is 50% in the population (the prevalence). In addition we know that the frequency of a specific feature in a bone (let's say a feature of the pubic symphisis) is higher in males of a certain age. Another way of saying this is that we know what the probability is of correctly ascertaining that a pubic symphisis belongs to a male when in fact the pubic symphisis belongs to males (sensitivity). We also know the probability of incorrectly diagnosing a pubic symphisis as female when the symphisis belongs to a male skeleton (specificity). With this information we can ascertain the probability of correctly determining that the pubic symphisis belongs to a male when the skeletal remains belong to a male. A Bayesian approach to probability was used to provide age estimates in modern American populations from the clavicle (Langley-Shirley and Jantz 2010). A Bayesian approach was also applied using several traits to identify one set of human remains. Specifically, Steadman *et al.* (2006) obtained separate estimates for identifying an individual from several traits (age, sex, stature, etc.). By multiplying these estimates the authors obtained a final probability which was three million times more likely to be obtained if the identification is correct than if the identification is incorrect. Clearly, this allows the forensic anthropologist to have a much more solid footing in the court room (Steadman *et al.* 2006). At the same time, it should be noted that Thevissen *et al.* (2010) report that a Bayesian approach produces more meaningful prediction intervals for aging individuals based on the development of the third molar, but that the Bayesian prediction does not "strongly outperform the classical approaches."

The second area in biological anthropology in which Bayes' theorem is contributing much is evolutionary studies of demographic history. Since a Bayesian approach to probability permits the researcher to use additional information to compute probabilities, it allows for a more complex modeling of the probabilities of events such as bottlenecks or expansions. By more complex modeling we mean the incorporation of multiple events such as the direction of gene flow, the speed by which two populations begin to separate, etc. For an excellent review of the use of Approximate Bayesian Computation (ABC) in evolutionary demographic studies, the reader is encouraged to review a piece by Bertorelle and collaborators (Bertorelle *et al.* 2010). As Bertorelle *et al.* note, a Bayesian statistical approach allows researchers to use prior information and to approximate posterior distributions assuming very complex models. An exciting application of the ABC approach in evolutionary studies is found in Wegmann and Excoffier (2010), who present a compelling revision of the recent evolutionary history of chimpanzees. This history shows that the central-African chimpanzees are the oldest subspecies, that the divergence time between the chimpanzee and bonobo species is only 0.9–1.3 million years ago, and that this divergence was very slow. This is a more complex view of the evolution of chimpanzees than would be possible to achieve if the researchers had only compared pairs of chimpanzee populations at a time.

3.6 The probability distribution of continuous variables

The process of computation of expected frequencies of outcomes can also be applied to a sample of continuous quantitative data. However, continuous numeric data present a complication not encountered in either qualitative or discontinuous numeric variables: continuous numeric data can take an infinite number of possible values. Indeed, the number of categories recorded in a continuous data set is ultimately dependent on the accuracy of the measuring tool. Let us illustrate this complication with an example in which I am asking you to take a leap of faith. Let us accept that we are able to know the height (in fractions of centimeters) of 20 subjects with the degree of accuracy shown in the first part of Table 3.4 (ungrouped data).

If we had employed a measuring scale such as those used in a doctor's office, we would have probably measured the subjects to the nearest centimeter, and we would have grouped them in 1-cm intervals, as shown in the second part of Table 3.4. Afterwards, we could proceed to group the data into 6-cm intervals to achieve more economy, as shown in the third part of the table.

Let us use the grouped and ungrouped data to compute the probability associated with events in this data set. First, let us use the 6-cm intervals. In our sample, what is the probability of finding someone whose height is over 147.500 and under 153.500? The probability is 15%. What is the probability of finding someone whose height is equal to/over 171.500 and under 177.500? The probability is 20%. Now let us use the 1-cm intervals. What is the probability of finding someone whose height is between 149 and 150? It is 5%. What is the probability of finding someone who is between 152 and 153? The probability is 0. What is the probability of finding someone who is between 156 and 157? It is 10%. Now let us look at the raw data. I asked you to take a leap of faith and accept that we knew the subjects' heights to this degree of accuracy. If we had used more common measuring tools we would have assigned subjects #6 and 7 in one category,

Table 3.4 The height (in centimeters) of 20 subjects. Ungrouped and grouped data. $\mu = 165$ cm, $\sigma = 11$.

ID	Height	Frequency	Percent	Cumulative frequency	Cumulative percent
1.	148.32040997	1	5.00	1	5.00
2.	149.19989617	1	5.00	2	10.00
3.	150.02151699	1	5.00	3	15.00
4.	153.91587449	1	5.00	4	20.00
5.	155.01365793	1	5.00	5	25.00
6.	156.89438989	1	5.00	6	30.00
7.	156.98606828	1	5.00	7	35.00
8.	158.63105665	1	5.00	8	40.00
9.	161.98573111	1	5.00	9	45.00
10.	163.10552002	1	5.00	10	50.00
11.	165.19325235	1	5.00	11	55.00
12.	169.96747687	1	5.00	12	60.00
13.	171.03060146	1	5.00	13	65.00
14.	171.84736183	1	5.00	14	70.00
15.	172.31853073	1	5.00	15	75.00
16.	176.83820614	1	5.00	16	80.00
17.	176.89106153	1	5.00	17	85.00
18.	179.2047241	1	5.00	18	90.00
19.	180.42990588	1	5.00	19	95.00
20.	181.05195193	1	5.00	20	100.00
Data grouped in 1-cm intervals (17 groups)					
148 <height< 149.00		1	5.00	1	5.00
149.00 ≤ height < 150.00		1	5.00	2	10.00
150.00 ≤ height < 151.00		1	5.00	3	15.00
153.00 ≤ height < 154.00		1	5.00	4	20.00
155.00 ≤ height < 156.00		1	5.00	5	25.00
156.00 ≤ height < 157.00		2	10.00	7	35.00
158.00 ≤ height < 159.00		1	5.00	8	40.00
161.00 ≤ height < 162.00		1	5.00	9	45.00
163.00 ≤ height < 164.00		1	5.00	10	50.00
165.00 ≤ height < 166.00		1	5.00	11	55.00
169.00 ≤ height < 170.00		1	5.00	12	60.00
171.00 ≤ height < 172.00		2	10.00	14	70.00
172.00 ≤ height < 173.00		1	5.00	15	75.00
176.00 ≤ height < 177.00		2	10.00	17	85.00
179.00 ≤ height < 180.00		1	5.00	18	90.00
180.00 ≤ height < 181.00		1	5.00	19	95.00
181.00 ≤ height < 182.00		1	5.00	20	100.00
Grouped data in 6-cm intervals (six groups)					
147.50 ≤ height < 153.50		3	15.00	3	15.00
153.50 ≤ height < 159.50		5	25.00	8	40.00
159.50 ≤ height < 165.50		3	15.00	11	55.00
165.50 ≤ height < 171.50		2	10.00	13	65.00
171.50 ≤ height < 177.50		4	20.00	17	85.00
177.50 ≤ height < 183.50		3	15.00	20	100.00

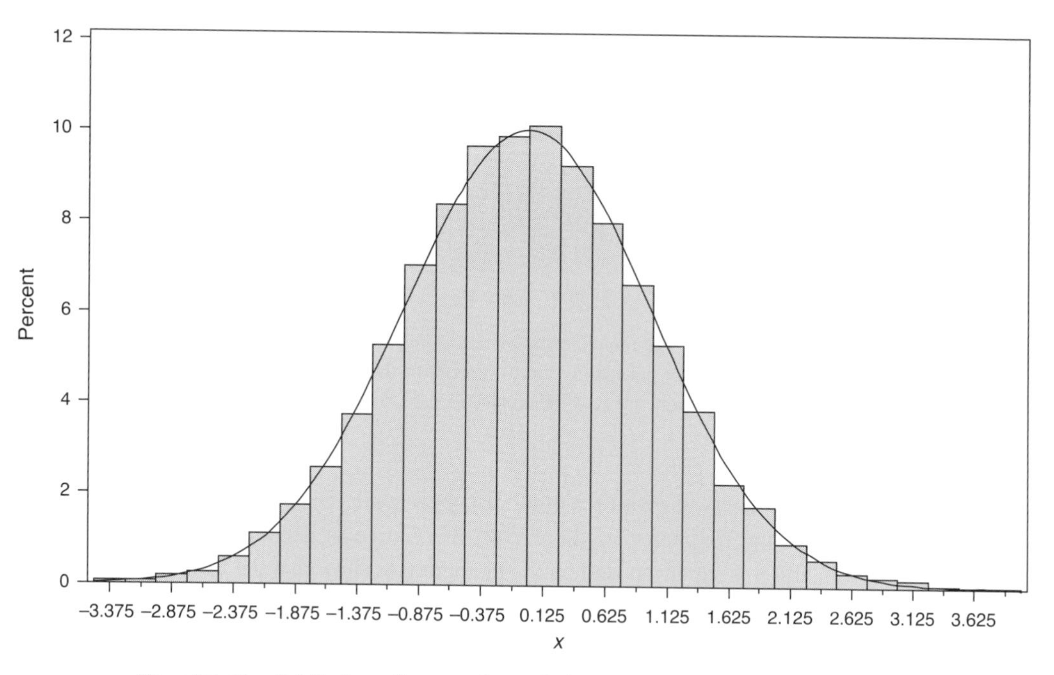

Figure 3.1 The distribution of a normal population (the histogram) with a standard normal distribution superimposed.

since their height is 156.89438989 and 156.98606828 respectively. The difficulty with computing the probability associated with events in continuous data sets is that there is an infinite number of possible events in our sample space, depending on the accuracy of the measuring tool.

Since we have better things to do with our lives than finding the probability associated with events that we never finish measuring, we define our events as intervals, which can be as small as we wish. For example:

If we measure our subject as	We can place him in this interval
148.5	148.495–148.505
148.55	148.5495–148.5505
148.555	148.55495–148.55505

Fortunately, finding the probability associated with events is greatly facilitated by the fact that many variables in nature *have* a **normal distribution**. When we say that a variable has a normal distribution, we mean to say that if we plot with a histogram all of the events in its sample space, the plot will approximate the look of a smooth bell. The narrower the intervals, the more the histogram will look like the bell. The bell which is the **standard normal distribution** is described mathematically by a function which requires more mathematical background than is assumed for this book, so the function is not presented here. Figure 3.1 shows a normally distributed variable (where the events

4:10 4:11 5:0 5:1 5:2 5:3 5:4 5:5 5:6 5:7 5:8 5:9 5:10 5:11 6:0 6:1 6:2

Figure 3.2 A group of 175 humans distributed according to their height. I wish to acknowledge the *Journal of Heredity* for permitting me to reproduce the picture shown in this figure, which appeared in "Corn and men" by A. F. Blakeslee (1914), **5** (11), 512.

are defined by intervals) with a normal distribution curve superimposed. The histogram itself shows the frequency of the events. The curve is derived by the mathematical function, and it is superimposed on the histogram to show that the former approximates the latter.

Statisticians and biologists have repeatedly shown that a large number of continuous numerical variables are normally distributed. Height is an excellent example of a normally distributed variable, as can be seen by Figure 3.2. The picture shows a group of 175 humans, distributed according to their height. The frequency of subjects is highest in the middle range of the distribution, and lowest in the two tails.

A normally distributed variable is said to be described by a **normal probability distribution**, which has the following attributes:

1. *Symmetry*: the curve is symmetric. That is, 50% of all observations are above and 50% below the mean (μ). Please note that this is the case independently of the value of the mean.
2. *Frequency of outcomes*: the line delimiting the "bell" actually tells us how frequent a particular outcome is. In other words, the higher the line associated with an outcome, the more likely the outcome is to occur. It is obvious that the outcome with the highest frequency is the mean (μ) and that the events with the lowest frequencies occur at both tails, where the line approaches **asymptotically** the horizontal axis without actually touching it. This is clearly exemplified by Figure 3.2: most individuals are close to the average height, and few are extremely short or extremely tall. But researchers cannot be sure that they have measured the absolutely tallest or shortest person in this or any other population. Thus, when we say that a variable is normally distributed, we say that most of the data take values close to the mean, with few towards the tails. Another way of saying this is that the probability associated with events is highest at the mean and that the probability of events decreases as we move away from the mean. We never say that the probability of finding a rare event is 0, but we may say that the probability of finding a very rare event approaches 0. I hope you can see how this type of statements stems from what we discussed in the introduction to this chapter about the fact that science produces probabilistic statements, not certainties.

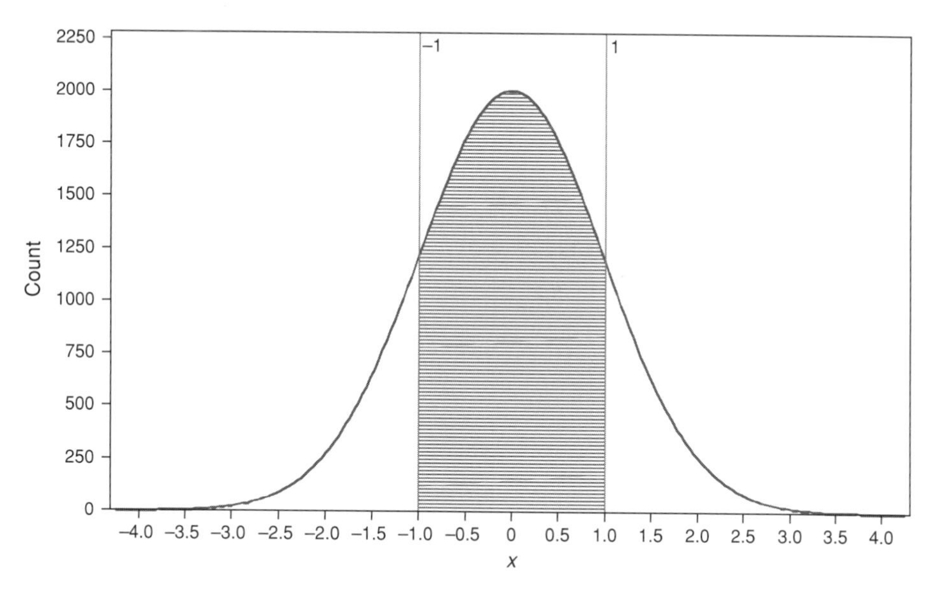

Figure 3.3 Area under the normal curve between one standard deviation on both sides of the mean.

3. *Percentage of items under the curve*: just like we know that 50% of the distribution is found to the right, and 50% to the left of the mean, we also know how items in each of the two halves are distributed within the normal curve. That is, we know that 34.135% of all items is found between the mean and the first standard deviation to its left, and 34.135% between the mean and the first standard deviation to its right. Thus, 68.27% (34.135 + 34.135) of all the population is found between the first negative and the first positive standard deviation (Figure 3.3).

We also know that 13.59% of outcomes is found between the first and second standard deviations. If this 13.59% is added to the 34.135% found between the mean and the first standard deviation, then 47.725% (13.59 + 34.135 = 47.725) of all the population is between the mean and the second standard deviation. Since the distribution is symmetrical, then 95.45% (47.725 + 47.725) of all outcomes is found between the mean and the two standard deviations on both sides (Figure 3.4).

Finally, we know that 2.14% of all outcomes is found between the second and third standard deviations, which when added to the 47.725% of outcomes between the mean and the second standard deviation adds up to 49.865% (2.14 + 47.725 = 49.865) on one side. Since the proportion is the same on both sides of the mean then 99.73% (49.865 + 49.865 = 99.73%) of all outcomes is found between the mean and the third standard deviation on both sides (Figure 3.5).

The reader can see now why the curve never actually touches the horizontal axis: 0.27 % of all outcomes (100 − 99.73) are found to the right and left of both third standard deviations. We can summarize this information as follows:

$\mu \pm \sigma$ contains 68.27% of all variants
$\mu \pm 2\sigma$ contains 95.45% of all variants
$\mu \pm 3\sigma$ contains 99.73% of all variants

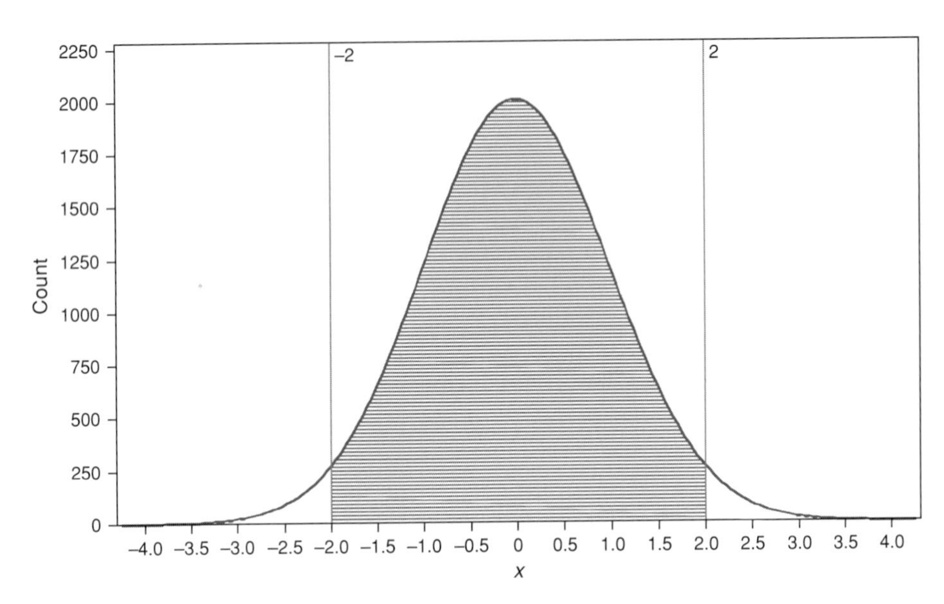

Figure 3.4 Area under the normal curve between two standard deviation on both sides of the mean.

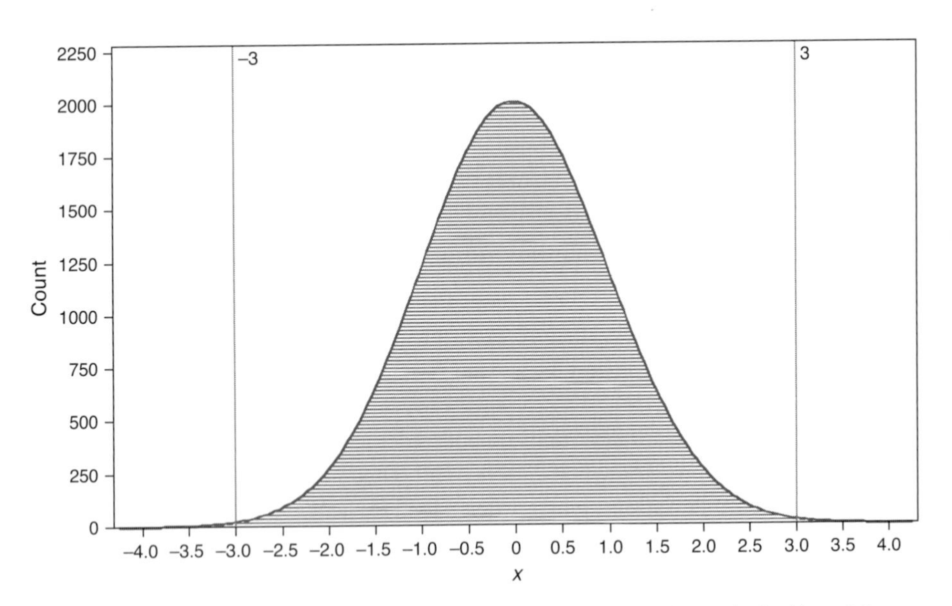

Figure 3.5 Area under the normal curve between three standard deviation on both sides of the mean.

4. *The probability associated with specific events in a continuous distribution*: recall how we estimated the probability of an event occurring in a qualitative data set. We determined that the frequency of the simple-stamped ceramic type in test unit A in Depot Creek Mound was $43/318 = 0.135$. We used this frequency to predict the

probability of finding this ceramic type, if we were to sample from the same population. The probability of finding events in a normal distribution rests on the same principle and can be intuitively grasped by looking at Figure 3.1. The reader can determine the frequency of an outcome by locating it in the horizontal line, then drawing a straight line up to the curve, and finding out the frequency pinpointed by the curve. But this is too cumbersome a method to find probabilities. Instead, we do the following: (1) We transform a continuous data distribution into a **standardized normal distribution** whose $\mu = 0, \sigma = 1$. (2) The probability associated with specific outcomes has already been computed for such a distribution, and is listed in the **Unit Normal Table**. The table can readily answer questions such as we were asking previously with our sample of 20 subjects whose heights we measured. We can define an interval (as broad as 6 or as narrow as 1 cm), and ask what the probability is of sampling a subject within those limits. Thus, we are going to do the following: (1) Using the raw data, compute μ and σ. (2) Transform the outcome whose probability of occurrence we wish to know into a "z score," or a "**standardized normal deviate**," and use the Unit Normal Table to estimate the probability associated with this outcome. This entire process is described in the section below.

3.6.1 *z* scores and the standard normal distribution (SND)

As we saw in the previous section, the process of computing the probability associated with an event for a continuous variable can be quite cumbersome. As we also mentioned, many continuous numerical variables have a distribution of outcomes which closely follows a bell curve. Indeed, Figure 3.1 shows the distribution of a variable (the histogram) with a line superimposed on it, where the line is known as the standardized normal distribution. The **standardized normal distribution** is a normal distribution with $\mu = 0$ and $\sigma = 1$. It is also the lifesaver which allows us to easily compute the probability associated with outcomes in a continuous quantitative variable whose mean and standard deviation can have any value. All we need to do is to standardize the outcome of our interest into a *z* score, and use the table which shows the areas under the normal curve associated with *z* scores (Table 3.5). A *z* **score** is very easily computed by subtracting the value of an observation from the population mean and dividing by the standard deviation of the population from which the observation was sampled. If we transform our entire data set into *z* scores we have standardized or centered or normalized the data. You will see these three different terms (**standardizing, normalizing, centering**) used interchangeably in the literature to refer to the same idea: transforming a data set into a set of *z* scores by the procedure described in this section.

The computation of a *z* score requires knowledge of population parameters. Since we do not usually have this information about the population, the use of *z* scores in real-life hypothesis testing is not too broad. However, most statistical tests covered in this book follow the principles we use when computing a *z* score to determine the probability associated with an event. It is for that reason that a solid understanding of *z* scores is fundamental. The formula for a *z* score is shown in formula 3.6.

Formula 3.6

Formula for the computation of a z score

$$z = \frac{y_i - \mu}{\sigma}$$

Here, y_i is the outcome whose probability of occurrence we wish to ascertain. For the time being we will assume that the data collected from the 20 study participants shown in Table 3.4 constitute a population for which we know its mean and standard deviation. What is the probability of sampling a subject whose height is 158.73? We are not ready to answer that question just yet, but we begin by converting the observation into a z score. Since our population has a mean height of $\mu = 165$ cm and a standard deviation of $\sigma = 11$, the z score for a subject whose height is 158.73 cm is $z = \frac{158.73-165}{11} = -0.57$.

Now we can learn how to use the table with **areas under the normal curve**, which is partly reproduced in Table 3.5 (all z scores we will use in this chapter are reproduced in this table). The table has three columns: **column A** lists the z scores, **column B** the area of the curve between the mean and the z score, and **column C** the area of the curve beyond the z score. With column B you can tell what percentage of the entire distribution lies between the mean and the z score, while with column C you can tell what percentage of the entire distribution is beyond (or has more extreme values than) our z score. These areas are illustrated in Figures 3.6 and 3.7. Please find the z score of 1.96 in Table 3.5, and note that its C column has a value of 0.025. This z score is important because if we consider the area beyond it on both tails, it delimits the difference between 95% and 5% of the distribution of z scores (where 5% is equally distributed on both tails).

Before we ask how likely we are to find a subject whose height is 158.73, we will ask some preliminary questions. The first thing we need to do is to locate the z score associated with that outcome, namely, $z = -0.57$. A quick view of column A shows that there are no negative scores. This is because a normal distribution is symmetrical, so we can answer all our questions by looking at the positive side of the distribution. Thus, we find the z score 0.57. With columns B and C we can answer two simple questions: what is the proportion of the distribution between the mean and a subject whose height is 158.73 (and whose z score is -0.57)? Column B answers that question: 0.2157%. What is the probability of finding a subject who is as short, or shorter than our subject? Column C answers that question: 0.2843%. When working with z scores, it is useful to draw the normal distribution, marking the area of enquiry, as done in the subsequent paragraphs.

Now let us ask a question that involves an interval: what is the probability of sampling a subject whose height is between 158.73 and 171.27? Since 158.73 is to the left, and 171.27 is to the right of the mean, we need to add the areas under B column for both z scores. Thus $z_1 = \frac{158.73-165}{11} = -0.57$, $z_2 = \frac{171.27-165}{11} = 0.57$. The area under the curve between the mean and the z score is 0.2157 and it is shown in Figure 3.8. Therefore, the

Table 3.5 Areas under the normal curve. Selected *z* scores are shown.

Column A (z scores)	Column B (area between mean and z)	Column C (area beyond z)
	⋮	
0.18	0.0714	0.4286
0.25	0.0987	0.4013
0.27	0.1064	0.3936
0.28	0.1103	0.3897
	⋮	
0.45	0.1736	0.3264
	⋮	
0.56	0.2123	0.2877
0.57	0.2157	0.2843
0.58	0.2190	0.2810
	⋮	
0.85	0.3023	0.1977
0.86	0.3051	0.1949
	⋮	
0.90	0.3159	0.1841
	⋮	
1.00	0.3413	0.1587
	⋮	
1.29	0.4015	0.0985
1.30	0.4032	0.0968
	⋮	
1.43	0.4236	0.0764
	⋮	
1.57	0.4418	0.0582
	⋮	
1.65	0.4505	0.0495
1.67	0.4525	0.0475
	⋮	
1.73	0.4582	0.0418
1.74	0.4591	0.0409
	⋮	
1.96	0. 4750	0.0250
	⋮	
2.14	0.4838	0.0162
2.17	0.4850	0.0150
2.18	0.4854	0.0146
	⋮	

(*cont.*)

Table 3.5 (*cont.*)

Column A (z scores)	Column B (area between mean and z)	Column C (area beyond z)
2.50	0.4938	0.0062
2.55	0.4946	0.0054
2.56	0.4948	0.0052
2.57	0.4949	0.0051
	⋮	
2.86	0.4979	0.0021
2.93	0.4983	0.0017
2.94	0.4984	0.0016
	⋮	
3.90	0.49995	0.00005
4.00	0.49997	0.00003

Column A lists the scores ($z = \frac{y-u}{\sigma}$).
Column B shows the proportion of the area between the mean and the z score value.
Column C shows the proportion of the area beyond the z score.

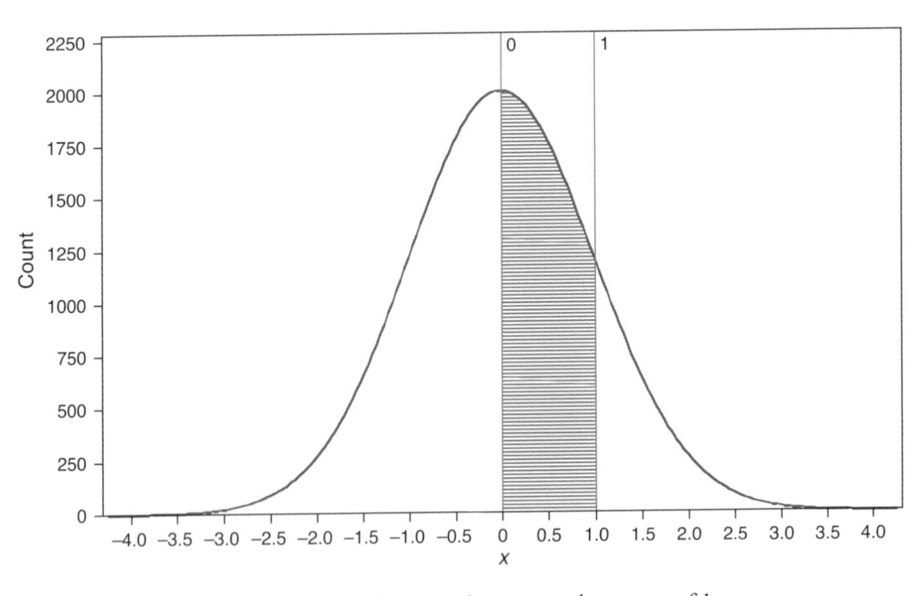

Figure 3.6 Area under the normal curve between the mean and a z score of 1.

area of the curve between both z scores (and the probability of sampling a subject who is between 158.73 and 171.27) is $0.2157 + 0.2157 = 0.4314$.

Let us ask another question that involves an interval: what is the probability of finding a subject whose height is between 158.73 and 160? In this case both observations are on the same side of the mean, so we cannot add the areas under both columns B as we

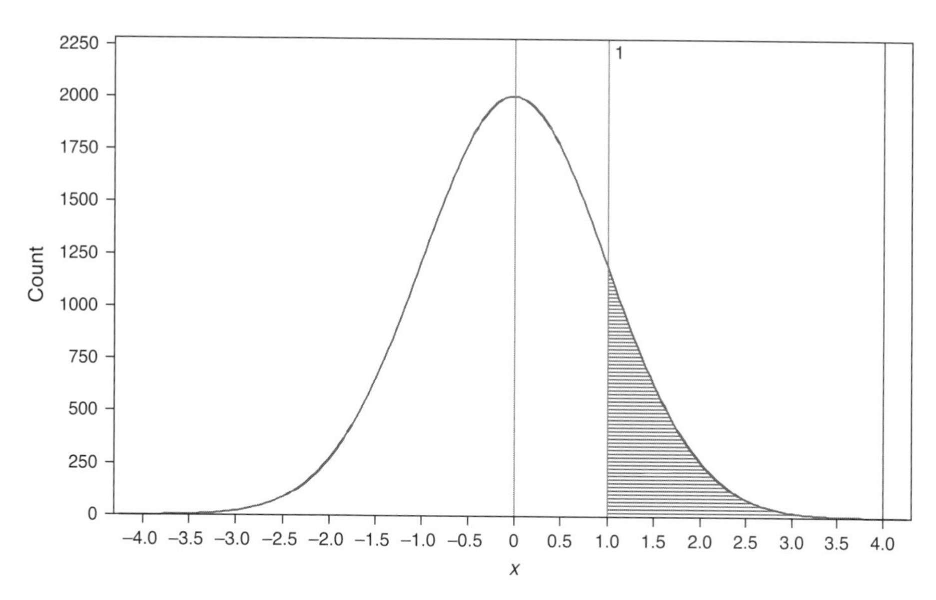

Figure 3.7 Area under the normal curve beyond a z score of 1.

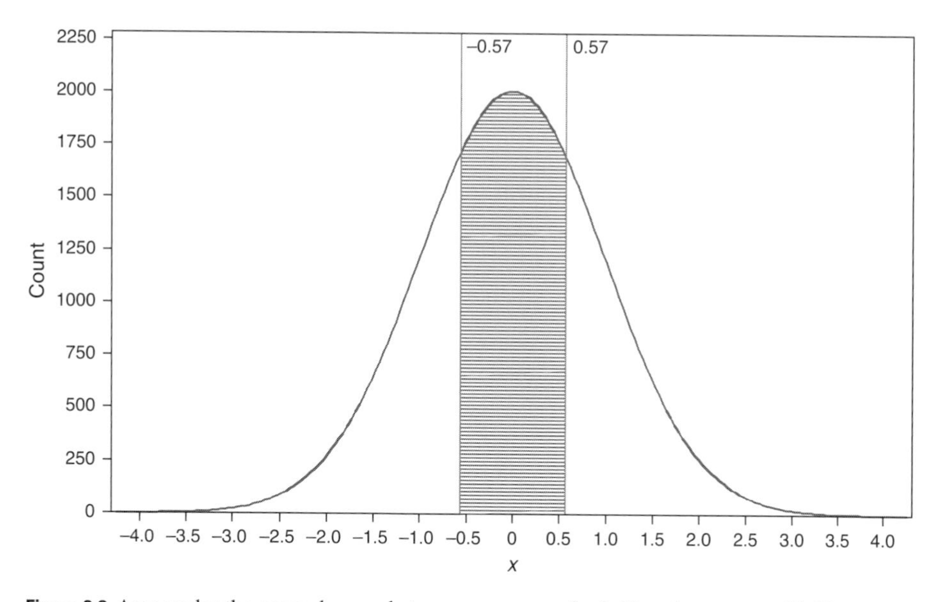

Figure 3.8 Area under the normal curve between a z score of -0.57 and a z score of 0.57.

did in the previous paragraph. Instead, we subtract from the larger area the smaller one. We begin by computing both z scores. $z_1 = \frac{158.73-165}{11} = -0.57$, $z_2 = \frac{160-165}{11} = -0.45$. The area between the mean and the first z score is 0.2157, while the area for the second z score is 0.1736. The difference between them is the probability of finding a subject between these two z scores. Thus, $0.2157 - 0.1736 = 0.0421$ is the probability of finding someone between 158.73 and 160. This area is shown in Figure 3.9.

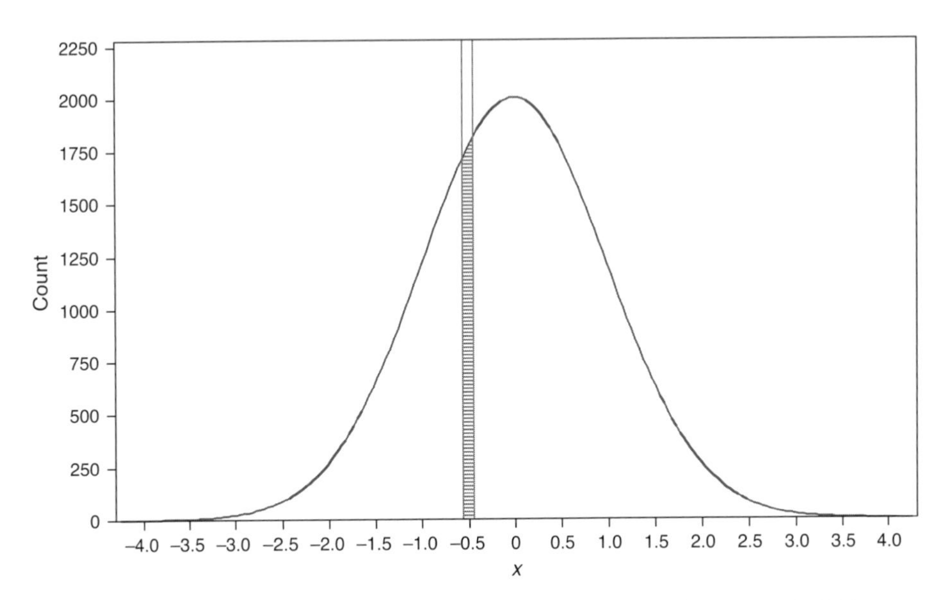

Figure 3.9 Area under the normal curve between a z score of -0.57 and a z score of -0.45.

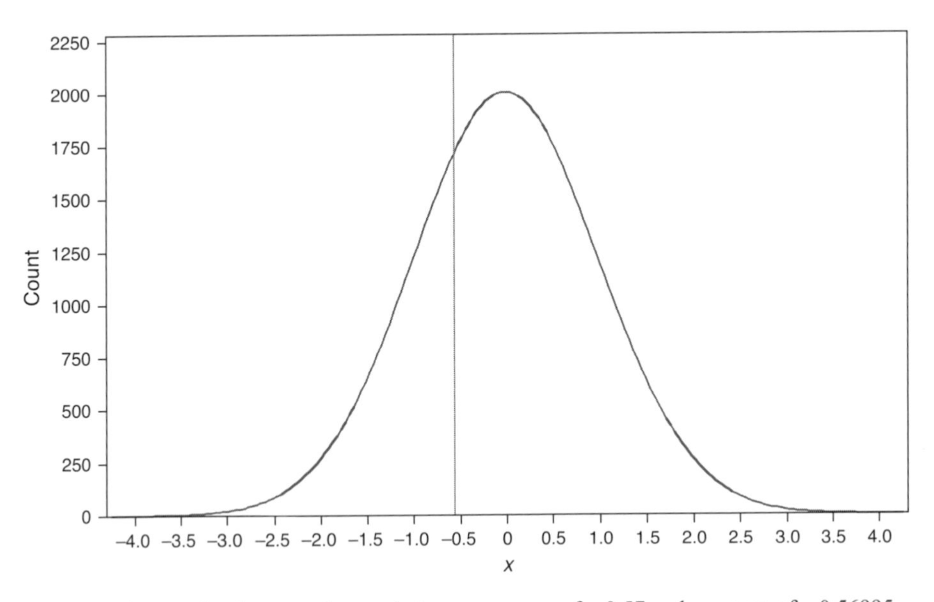

Figure 3.10 Area under the normal curve between a z score of -0.57 and a z score of -0.56995.

Finally, to answer what the probability is of finding a subject whose height is 158.73 in this population, we need to think of the observation as being in an interval. We can define the interval as 158.7295–158.7305 and place 158.73 comfortably in it (we are doing here what we did before, when we were reducing the size of the interval which contained an observation). So, what is the probability of sampling someone whose height falls within 158.7295–158.7305? We need to compute two z scores, both of which are on the same side of the curve since both are below the mean. $z_1 = \frac{158.7295-165}{11} = -0.57$, $z_2 = \frac{158.7305-165}{11} = -0.56995$. The area between the mean and the first z score is 0.2157 while the area for the second z score is 0.2122 (note: this value was obtained by a computer program because the table of areas under the curve does not have this precise z score). If we subtract the latter from the former, then $0.2157 - 0.2122 = 0.0035$. The probability of finding such a precise measure is much smaller than the probability associated with a larger interval, and it is shown in Figure 3.10.

Practice problem 3.5

Let us work now with a population of females with a mean stature of $\mu = 160$ and a standard deviation of $\sigma = 7$ to answer the following questions:

1. What is the probability of finding a female whose height falls between 154 and 167 cm? Note that these two observations are on opposite sides of the mean (which is 160). Thus, we need to find the area between the mean and the z scores, which is provided by column B, and sum them. First, we compute the z scores:

$$z = \frac{154 - 160}{7} = -0.86, \quad \text{and } z = \frac{167 - 160}{7} = 1.0.$$

Then we find the values of column B for both z scores: for $z = -0.86$ the area is 0.3051, and for $z = 1$ the area is 0.3413. We sum them for the answer: the probability of finding a female whose height falls somewhere between 154 and 167 is $0.3051 + 0.3413 = 0.6464$. This is illustrated in Figure 3.11.

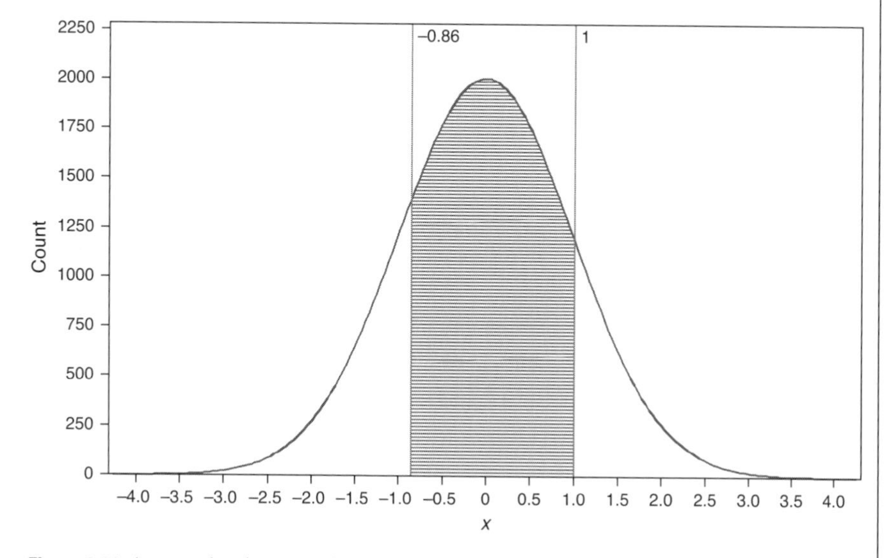

Figure 3.11 Area under the normal curve between a z score of -0.86 and a z score of 1.0.

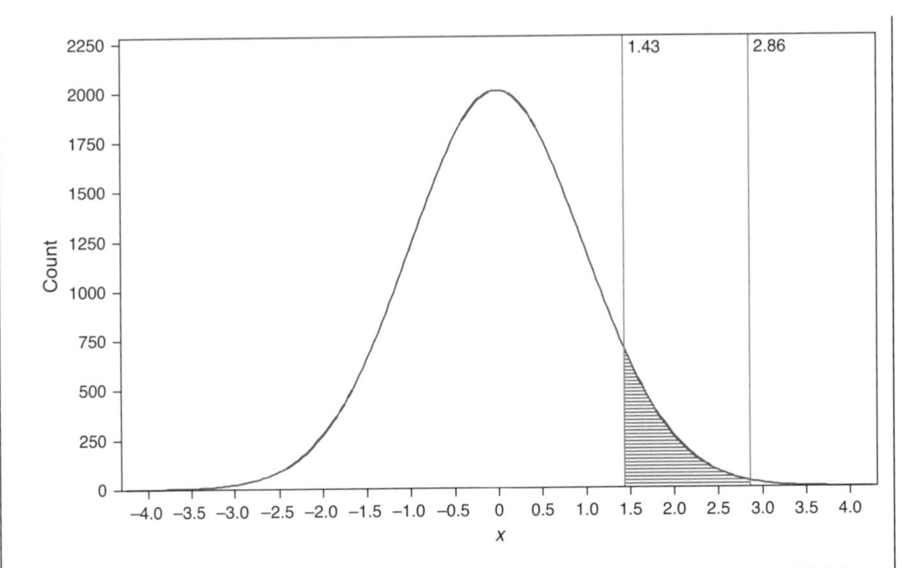

Figure 3.12 Area under the normal curve between a z score of 1.43 and a z score of 2.86.

2. What is the probability of sampling a female whose height is between 170 and 180 cm? Both outcomes are on one side of the distribution, and we want to find the probability delimited by both z scores. To find it, we need to subtract the smaller column B value from the larger one (see Figure 3.12). We proceed as follows:

z scores	Column B	Answer
$z = \frac{170-160}{7} = 1.43.$	0.4236	
$z = \frac{180-160}{7} = 2.86.$	0.4979	$0.4979 - 0.4236 = 0.0743$

3. Let us now narrow our inquiry: What is the probability of sampling a female whose height is between 170 and 175 cm? (See Figure 3.13.)

z scores	Column B	Answer
$z = \frac{170-160}{7} = 1.43.$	0.4236	
$z = \frac{175-160}{7} = 2.14.$	0.4838	$0.4838 - 0.4236 = 0.0602$

4. What is the probability of sampling a female whose height is between 170 and 171 cm? (See Figure 3.14.)

z scores	Column B	Answer
$z = \frac{170-160}{7} = 1.43.$	0.4236	
$z = \frac{171-160}{7} = 1.57.$	0.4418	$0.4418 - 0.4236 = 0.0182$

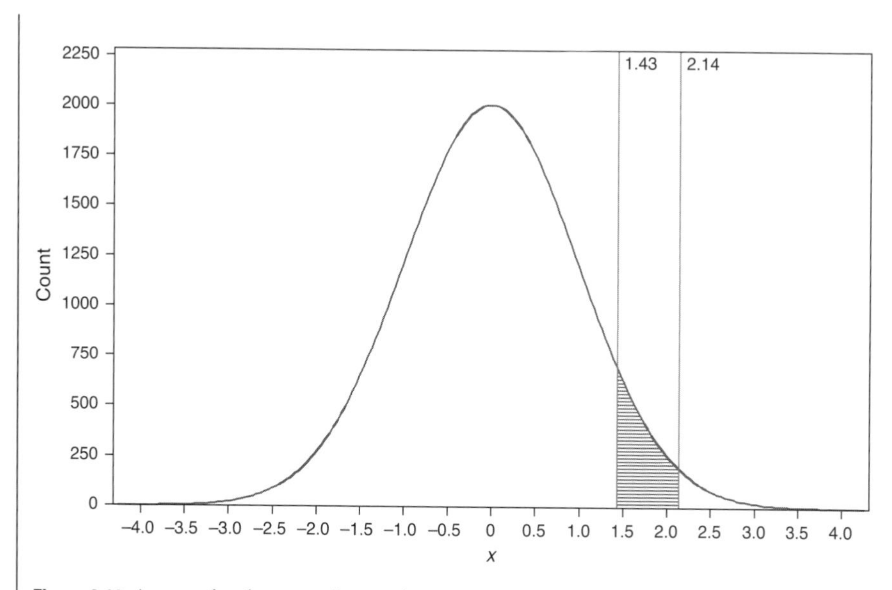

Figure 3.13 Area under the normal curve between a z score of 1.43 and a z score of 2.14.

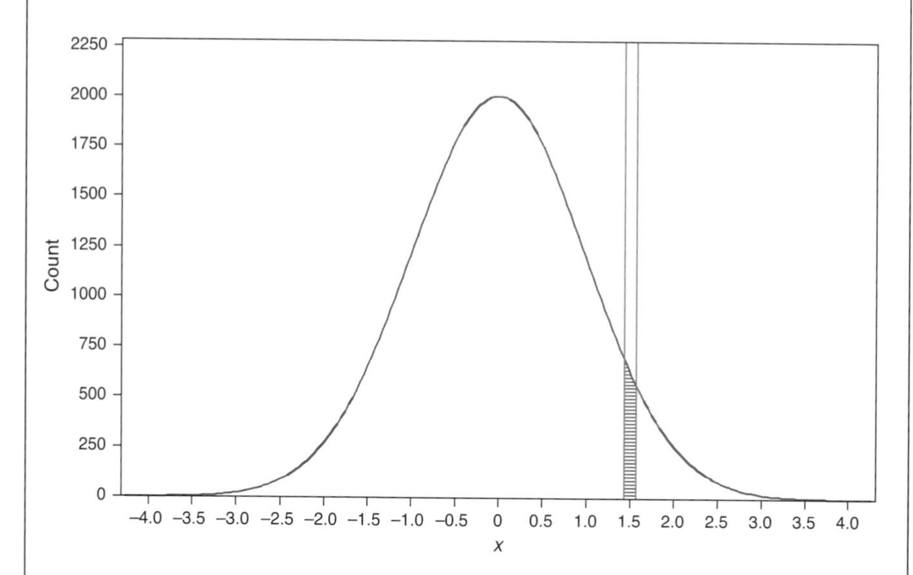

Figure 3.14 Area under the normal curve between a z score of 1.43 and a z score of 1.57.

This practice shows that the probability associated with events depends not only on how far away the event is from the mean (whether the z score is 1 or 4), but also on how the event itself is defined (170–171 or 170–175).

3.6.2 Percentile ranks and percentiles

We all have had to deal with percentiles. Graduate students take the GRE, and would much rather be in the 90th than in the 50th percentile. The proud parents of babies eagerly

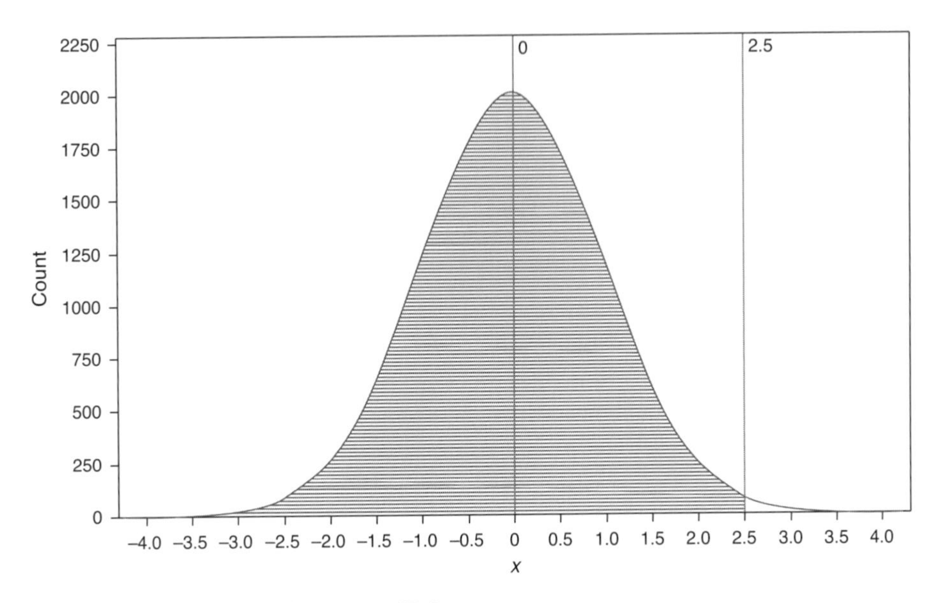

Figure 3.15 Percentile rank of a z score of 2.5.

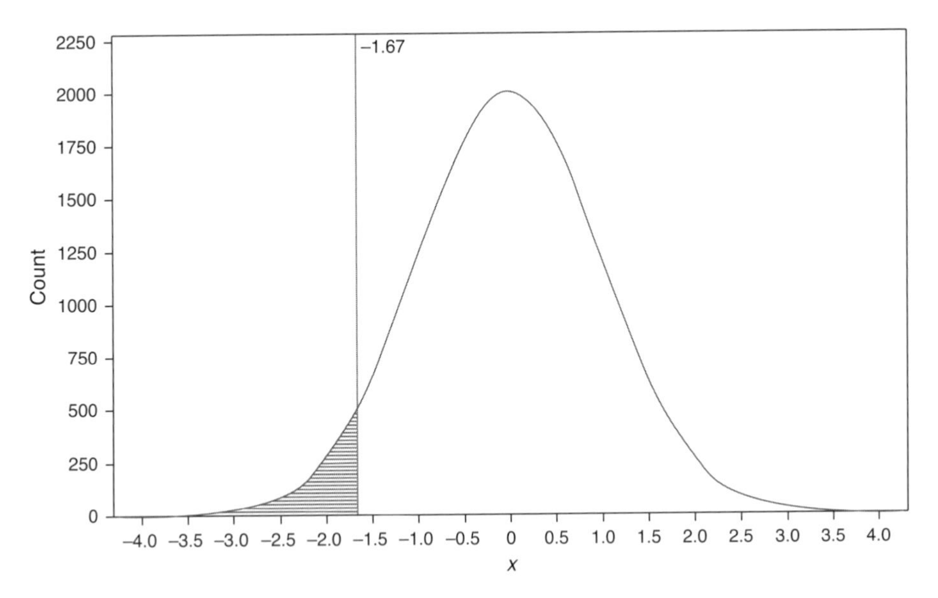

Figure 3.16 Percentile rank of a z score of -1.67.

await their offspring's monthly visit to find out what height and weight percentile the baby is in. Percentiles are indeed a popular use of z scores. The **percentile rank** is the percentage of individuals in the distribution with scores at or below that particular score. The specific score associated with a percentile rank is called a **percentile**. If we say that a test score falls in the 50th percentile, then it fell at the center of the distribution. If it is in the 60th percentile, 60% of the distribution is below it.

To find out in what percentile an observation falls, it is transformed into a z score. Let us say that in a population of test scores whose $\mu = 110$ and $\sigma = 12$ a student gets a score of 140. What percentile is associated with this datum? We compute a z score as always: $z = \frac{140-110}{12} = 2.5$. The test score is higher than the mean, so at least we know that it is higher than the 50th percentile. Then we look for the value of column B associated with $z = 2.5$, which is 0.4938. If we add that to the 50% to the left of the mean, we find out that the score in the $0.5 + 0.4938 = 99$th percentile (Figure 3.15).

What if the test score was, however, 90? The z score would be: $z = \frac{90-110}{12} = -1.67$. We need to use column C here, because we are looking for the proportion of the distribution with values under this grade. For $z = -1.67$, column c $= 0.0475$. Sadly, the score is in the 4th percentile.

Practice problem 3.6

Let us use the population shown in Table 3.4, whose $\mu = 165$ and $\sigma = 11$ cm. What is the percentile rank of someone who is 174.9 cm tall? We know that this person has a percentile rank over 50% since his height is greater than the mean. Therefore, we need to find out the value for column B and add it to 0.5. The z score is $z = \frac{174.9-165}{11} = 0.90$. The area under column B for a z score of 0.90 is 0.3159. If we add this to 0.5, then we find out that this person is in the 81st percentile rank. The percentile rank for this individual is $0.5 + 0.3159 = 0.8159$.

3.6.3 The probability distribution of sample means

This section deals with an application of probability in statistics that is more relevant to actual research. Let us say that a cultural anthropologist is investigating the age at which males marry in a population. The anthropologist ascertains that the mean sample age at marriage is 20 years. But, how representative is this $\bar{Y}$ from the population's μ? This section will cover the probability of obtaining a sample of size n with a specific $\bar{Y}$, *assuming* knowledge of the population's μ and σ. Obviously, this is an unrealistic situation, since we rarely have such population knowledge. However, probability deals with hypothetical situations, and for us to understand statistical reasoning, we need to understand the workings of probability in such hypothetical scenarios. In the next chapter we will make the transition to the "real world."

Let us continue our discussion with a pedagogical example of a population of only $N = 3$ subjects whose values are: $y_1 = 10$, $y_2 = 20$, and $y_3 = 30$. Let us now proceed to sample with replacement all possible samples of size $n = 2$, and compute for each sample its $\bar{Y}$, as well. Since we are sampling with replacement, the sample could consist of the same score twice. Starting with y_1, the samples could include y_1 and y_1, y_1 and y_2, and y_1 and y_3, and so forth. Table 3.6 shows all possible samples, their sample means, and a frequency distribution of the sample means. A bar graph of the sample means is shown in Figure 3.17. It should be apparent that if all possible samples of size n are

Table 3.6 The distribution of sample means of a known population.

Sample number	First observation	Second observation	$\bar{Y}$
1	10	10	10
2	10	20	15
3	10	30	20
4	20	10	15
5	20	20	20
6	20	30	25
7	30	10	20
8	30	20	25
9	30	30	30

The frequency distribution of the sample means is:

Sample means	Frequency (f)	Percent
10	1	$1/9 = 0.111$
15	2	$2/9 = 0.222$
20	3	$3/9 = 0.333$
25	2	$2/9 = 0.222$
30	1	$1/0 = 0.111$
$\mu = 20$	$\sum f = 9$	$0.999 \approx 1$

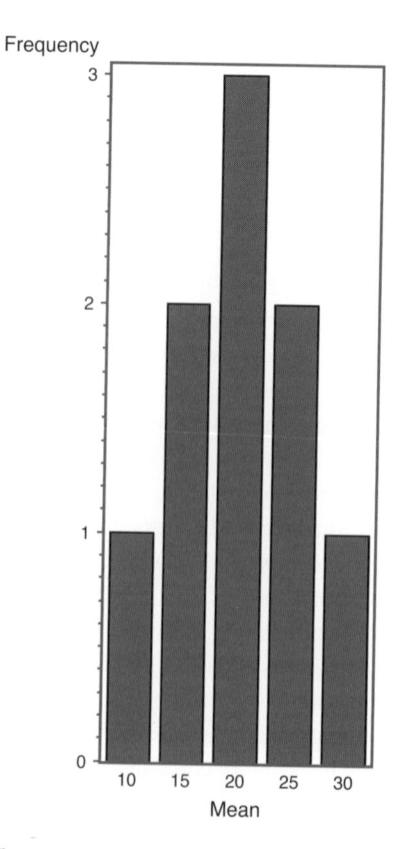

Figure 3.17 A histogram of the table means shown in Table 3.6.

sampled from a population, the sample means are approximately normally distributed. Notice that our population was most certainly not normal: it had a flat shape, with three subjects found at the same frequency (instead of a normal bell curve, with the highest frequency found at the mean).

The distribution of sample means is described by the **central limit theorem**. Succinctly, this theorem tells us that if we sample the sample means of a normally distributed population the distribution of sample means will be normally distributed as well. Moreover, if we sample from a non-normal distribution the distribution of sample means will be approximately normal. What is key is that the larger the sample size, the more normally distributed the sample means will be.

We can now go a step further: Given that we have a distribution of sample means, we can compute the probability associated with particular outcomes (sample means) as done previously in this chapter. Such probabilities are also shown in Table 3.6. Thus, what is the probability of obtaining a sample of $n = 2$ with a mean of $\bar{Y} = 10$? Since only one sample had this mean, the probability is easily computed by dividing $1/9 = 0.11$, where $\sum f = 9$ is the total number of samples. What is the probability of obtaining a sample of size $n = 2$ with a mean of $\bar{Y} = 20$? The probability of such an outcome is $p = 0.333$. Therefore, when we have total knowledge of a population, and can obtain all possible samples of size n, we can quantify the probability associated with obtaining a specific outcome or $\bar{Y}$. Imagine what a cumbersome task that would be if instead of having a population size N = 3, and instead of obtaining all possible samples of size $n = 2$, our population was of size N = 1000 and we wanted to know the probability distribution of samples with size $n = 30$. Thankfully, we do not engage in such an onerous task but rely on z scores again. Instead of computing the probability associated with specific outcomes as we did above, we can transform our sample mean into a z score, and use the table of areas under the normal curve to find the probability associated with such an outcome. The outcome of interest now is the sample mean $\bar{Y}$ instead of the observation y_i.

As the reader knows, to compute a z score we need to divide the difference between $\bar{Y}$ and μ by some kind of standard deviation. However, what we need in the present case is the standard deviation not of a sample of individuals, but a standard deviation of sample means. The value of the **standard deviation of the sample means** (also known as the **standard error of the means**) depends on two items: the population's standard deviation (σ, which is the standard deviation of the population's subjects), and the size of the sample. *Assuming knowledge of the parametric standard deviation*, the formula for computing the standard deviation of sample means is shown in formula 3.7.

Formula 3.7

The standard deviation/error of the means.

$$\sigma_{\bar{Y}} = \frac{\sigma}{\sqrt{n}}$$

An interesting aspect of the standard deviation of the means is that as sample size increases the scatter around the mean of the population of sample means decreases. If $\sigma = 8.16$ and $n = 2$, then $\sigma_{\bar{Y}} = \frac{8.16}{\sqrt{2}} = 5.77$. If, however, $n = 5$, then $\sigma_{\bar{Y}} = \frac{8.16}{\sqrt{5}} = 3.65$, and if $n = 10$, then $\sigma_{\bar{Y}} = \frac{8.16}{\sqrt{10}} = 2.58$. Clearly, if sample sizes are increased, the distribution of the means will be less scattered.

There is a good reason why the standard deviation of the means is referred to as the standard error of the means. Because the larger the value of $\sigma_{\bar{Y}}$, the more scattered the sample means will be from the population mean, the more error there will be in the estimation of the population parameter. Conversely, if the standard error of the means is small, the sample means will be closer to the population parameter, and the error in the estimation of the population parameter will be diminished. In this book we will use interchangeably the terms standard deviation of the means and the standard error of the means. Please note the difference in the notation between σ and $\sigma_{\bar{Y}}$. The former refers to the standard deviation of the observations in the population, and the latter refers to the standard deviation of means of samples of the same size, taken from the population.

We are now in a position to answer the question we asked at the beginning of this section: assuming knowledge of the population, what is the probability of obtaining a sample with a specified n and $\bar{Y}$? We will answer that question by computing a z score which uses $\sigma_{\bar{Y}}$ instead of σ. Thus, our z score formula is slightly changed to:

Formula 3.8

Formula for the z score of a sample mean.

$$z = \frac{\bar{Y} - \mu}{\frac{\sigma}{\sqrt{n}}} \quad \text{or} \quad z = \frac{\bar{Y} - \mu}{\sigma_{\bar{Y}}}$$

Practice problem 3.7

An anthropologist is interested in the size of homesteads in a rural county in the Mid West region of the USA. The anthropologist is interested in determining if land ownership is different between families long established in the region and families who are recent migrants and who tend to reside in a particular area of the county. Using county records, he establishes that the mean homestead size in the county is 397 acres with a standard deviation of 85. The anthropologist interviews 30 migrant male heads of households ($n = 30$) and asks how large each migrant family's holding is. From the interviews he determines that the mean farm size in the migrant community is 200 acres. What is the probability associated with obtaining such an outcome or anything more extreme in a population with a mean of 397 and a standard deviation of 85 acres?

$$\text{The } z \text{ score is: } z = \frac{200 - 397}{\frac{85}{\sqrt{30}}} = \frac{-197}{15.52} = -12.69$$

The probability associated with this z score approaches 0. It is highly unlikely to find a sample mean of 200 in a population with a μ of 397.

3.6.4 Is my bell shape normal?

It is easy to assume by default that a bell-shaped distribution is normal. This is, however, not always the case. The distribution of continuous variables may approximate a normal distribution but may suffer from significant deviations from a normal distribution. An obvious deviation is that of **skewness**, in which the data are "clustered" on one side (if they are clustered on the right side of the distribution, the data are skewed to the left – see Figure 3.18. If the data are clustered to the left then they are skewed to the right – see Figure 3.19. I always found this slightly confusing).

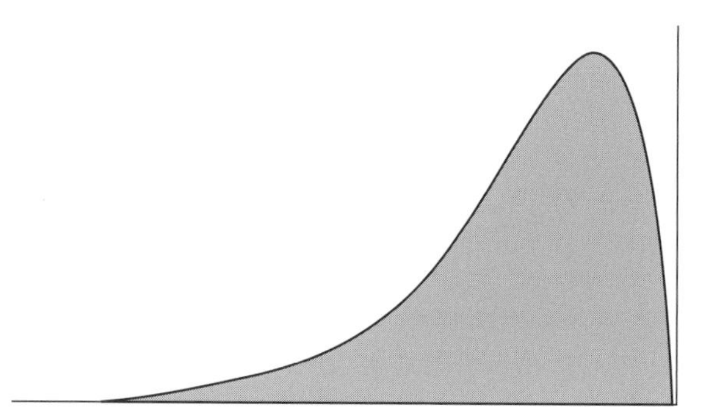

Figure 3.18 A data set skewed to the left.

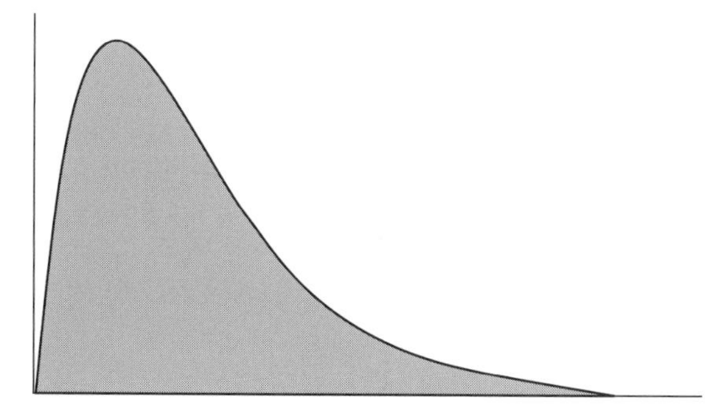

Figure 3.19 A data set skewed to the right.

Another deviation from normality is **kurtosis**, which describes the distribution of the data in the tails, the intermediate and central areas of the distribution. Some populations are too "peaked" in which case they are called **leptokurtic** –see Figure 3.20 – while

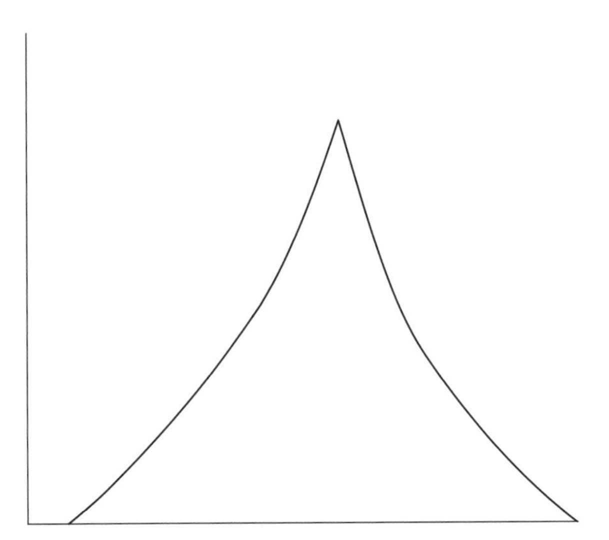

Figure 3.20 A leptokurtic distribution.

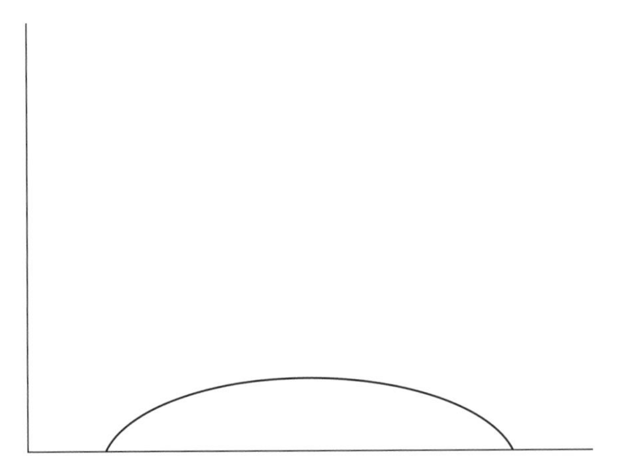

Figure 3.21 A platykurtic distribution.

others are **platykurtic** in which case they are too flat and do not asymptotically approach the X axis – see Figure 3.21. Finally, it is also possible for a distribution to be bimodal, in which case there are two areas of the distribution which are highly frequent – see Figure 3.22.

A normal distribution is assumed for many statistical tests, and we will learn how to test if a data set follows a normal distribution.

3.7 Chapter 3 key concepts

- In this chapter we dealt with hypothetical situations, in which we assumed knowledge of the population parameters. In the next chapter we will deal with "real-life"

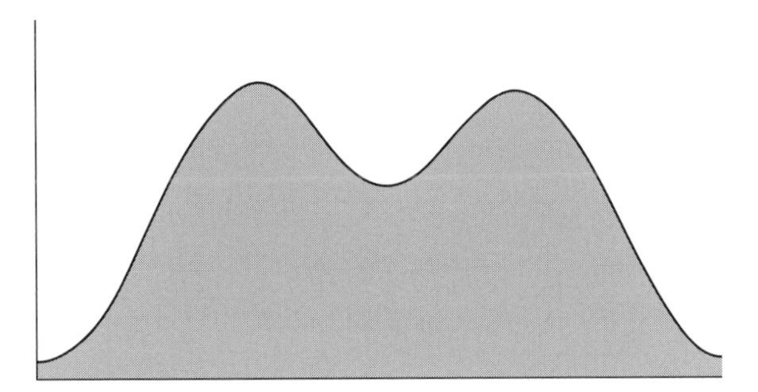

Figure 3.22 A bimodal distribution.

situations, in which we do not have knowledge of the population's parameters. Perhaps the most important concept of the entire chapter is that probability provides the foundation upon which we can make scientific statements. A person who says that there is "scientific proof for the occurrence of an event. . . " is actually saying that "the probability associated with the event is close to 1."

- Describe sampling with and without replacement.
- The probability distribution of qualitative variables.
- The probability distribution of discrete quantitative variables.
- When do we use the binomial distribution?
- When do we use the Poisson distribution?
- Explain what we mean when we say that there are an infinite number of outcomes in a continuous variable frequency distribution.
- What are the advantages of using a Bayesian approach to probability?
- z scores and areas under the normal curve.
- What are percentile ranks?
- How do z scores differ if we are dealing with a single event or a sample mean?

3.8 Computer resources

1. For the computation of Poisson distribution in PASW, you need to be very clear as to which variable is the Y (the counts, the projectile points, the number of new pneumonia cases) and the frequency or the number of times the Y_i was observed (how many square meters, how many days). After you enter both columns, you weight the Y by the frequencies at data → weight cases → choose the frequency. Then you go to transform → compute variable → Here you give a name to the new variable, say "Poisson." Choose PDF and non-central PDF and PDF Poisson. You will have to enter two quantities (cleverly highlighted by two question marks). The first is the frequency variable (the counts of projectile points) and the second one is the mean. Once you get the relative expected frequencies, you can easily obtain the absolute frequencies by declaring a new variable at transform → compute variable → Here

you give a name to the new variable, say absolute, and by using the "keyboard" in this window to ask PASW to multiply the relative frequencies by the total number of events or $\sum fY$.
2. All graphs of the normal curve were obtained with SAS (Proc capability).
3. There are many online "calculators" which allow you to obtain the area under the normal curve.

3.9 Chapter 3 exercises

1. Describe ways to ensure procurement of a random sample. If you choose not to use replacement, justify this in the context of a research situation.
2. Compute the expected genotype frequencies for the following hemoglobin data on a fictitious sample in a subtropical population:

HbA = 0.92	HbS = 0.08

3. A social marketing researcher has sampled a group of older men from a diverse community, to determine if they have been checked for prostate cancer. The data are in an Excel file called "prostate." You have been hired to do a follow-up study in the same community, in which you will sample a total of 80 subjects. Given the ethnicity of the males who were sampled already, construct a probability distribution, and compute the probability associated with each ethnic group. Given these probabilities, how many males (out of 80) would you expect to find in your new sample, who belong to each of these ethnic groups?

   Participant 1 Hispanic
   Participant 2 Hmong
   Participant 3 Hispanic
   Participant 4 African American
   Participant 5 African American
   Participant 6 Japanese
   Participant 7 Hispanic
   Participant 8 African American
   Participant 9 Hispanic
   Participant 10 African American
   Participant 11 Hispanic
   Participant 12 Greek
   Participant 13 African American
   Participant 14 Hispanic
   Participant 15 Chinese

Participant 16 African American
Participant 17 African American
Participant 18 Hispanic
Participant 19 African American
Participant 20 Samoan
Participant 21 Hispanic

4. A cultural anthropologist is interested in the economic role children play in the community he is studying. He realizes that one of the most important contributions children make is gathering a berry which appears to be scarce and which requires children to do a lot of walking around the settlement. The anthropologist decides to follow a child who is going out to gather the berry for several days, and records the number of berries the child collects every ½ hour. The anthropologist would like to know if the distribution of the berries follows a Poisson distribution. Y (our variable of interest) is the number of berries gathered, while the observed frequency is the number of ½ hour periods in which such numbers were observed (observed over several days). The data are shown below. Compute the absolute and relative expected Poisson frequencies. Do these data appear to follow a Poisson distribution?

Number of berries gathered	Intervals of ½ hour in which these berries were obtained
25	1
30	1
50	1
65	2
70	2
80	1
95	1

5. It is known that in a particular archaeological population, the mean number of linear enamel hypoplasias (defects in enamel development due to a metabolic insult to the organism) per individual is $\mu = 5$ with $\sigma = 2.0$. What is the probability of excavating the following:
   a. an individual with 1 or fewer lesions.
   b. an individual with between 3 and 6 lesions.
   c. an individual with between 2 and 4 lesions.
   d. an individual with 9 or more lesions.

6. It is known that the elderly women of a certain rural American population have a mean weight of $\mu = 150$ pounds, with a standard deviation of $\sigma = 10$. What is the probability of finding in such a population:
   a. a woman who weighs between 125 and 135 pounds?
   b. a woman who weighs between 140 and 160 pounds?
   c. a sample of size = 60 with a $\bar{Y} = 170$? What would your answer be if $n = 30$?
   d. a sample of size = 70 with a $\bar{Y} = 170$? What would your answer be if $n = 100$?

4 Hypothesis testing and estimation

4.1 Different approaches to hypothesis testing and estimation

Anthropologists have dealt numerous times with a situation in which a cultural manifestation achieves hegemony over others. For example, in many countries bio-medicine achieved the status of acceptable medicine while pushing other forms of medical care out to the fringes. In several historical cases a language became the *lingua franca* while displacing others. In the world of statistical analysis of quantitative data, the usual process of hypothesis testing which results in most of the hundreds of papers published in scientific journals is also a result of the hegemony of an ideology over others. It was not always done as it is today although most statistics textbooks (the first edition of this book included) do not acknowledge it.

In this chapter we cover two related topics: estimation and hypothesis testing, although the majority of the chapter concentrates on the latter. A good statistical background is necessary for an anthropological scientist because the mark of science is its ability to test and reject hypotheses. Hypothesis testing is at the core of what science does, and statistical analysis is a tool to accomplish it.

But how should hypothesis testing (and estimation) be done? What most statistics textbooks cover, and this textbook is not an exception, is what may be called "the classical significance testing approach." This approach includes the usual statistical tests expected in graduate-level theses and dissertations and in peer-reviewed journals. This is certainly a powerful and practical approach to the analysis of quantitative data. However, it is not the only one. In the following paragraphs I discuss superficially the development of the classical approach, and I mention two other approaches. Although this textbook follows the hegemony in statistical analysis, it at least acknowledges that it does.

4.1.1 The classical significance testing approach

The classical significance testing approach to data analysis is actually a combination of two different proposals from the early part of the 1900s. (1) Ronald Fisher championed the proposition of a single hypothesis and on determining the likelihood of finding the observed event or more extreme events if the hypothesis were true (Fisher 1993). This is very reminiscent of what we did in the previous chapter, when we found the probability associated with finding a very short individual. (2) Neyman and Pearson instead proposed that hypothesis testing should be a decision between a null and an alternative hypothesis, taking into consideration the "cost" associated with the probability of committing statistical errors in this decision (Neyman and Pearson 1933). These two proposals together (called a hybrid of contradictory proposals in Cohen 1992) have become the hegemony in statistical analyses of quantitative data despite frequent misunderstandings of its key concepts by practitioners, as pointed out by others (Pollard 1986; Towner and Luttbeg 2007). In this chapter I discuss estimation and hypothesis testing within the classical significance testing approach while pointing out common misunderstandings and misapplications by its users.

4.1.2 The maximum likelihood approach

The maximum likelihood approach to estimation is due to Fisher as well (Aldrich 1997). As we will see below, estimation is very closely related to hypothesis testing, and it is a necessary step in different processes for analyzing quantitative data. If we wish to estimate the parameters of a population, a **maximum likelihood estimator** of the parameters is one that maximizes the probability of the observed data being derived from a population with such parameters.

4.1.3 The Bayesian approach

The Bayesian approach to probability stems from an essay by the Reverend Thomas Bayes, read and published posthumously (Bayes 1991). As we learned in the previous chapter, this approach uses previous knowledge and prior estimates to calculate posterior estimates, which allow us to estimate the probability that a hypothesis is true. An excellent example of how we all engage in Bayesian probability is provided by Steadman *et al.* (2006), who note that the jury in a criminal case *is* engaging in Bayesian probability as it listens to witnesses and sees evidence, so that the jury is updating its estimates of posterior probability of the events discussed in the court room (Steadman *et al.* 2006). A very strong case for adopting a Bayesian approach to evaluation research is proposed by Pollard (1986). Readers interested in program evaluation (for NGOs or government programs, for example) are encouraged to read this source (Pollard 1986).

The rest of this textbook is written within the classical significance testing approach to studying quantitative data. Readers are referred to an excellent review of the three approaches within the field of human behavioral ecology (Towner and Luttbeg 2007).

4.2 Estimation

When we compute a sample statistic, we frequently wish to make statements about the population parameter estimated by the statistic. In this section, we discuss how such estimation should proceed.

In the first chapter we discussed two important terms: **descriptive** and **inferential statistics**. With the former we describe the sample, with the latter we make statements about the population from which we obtained the sample. In chapter two we learned to

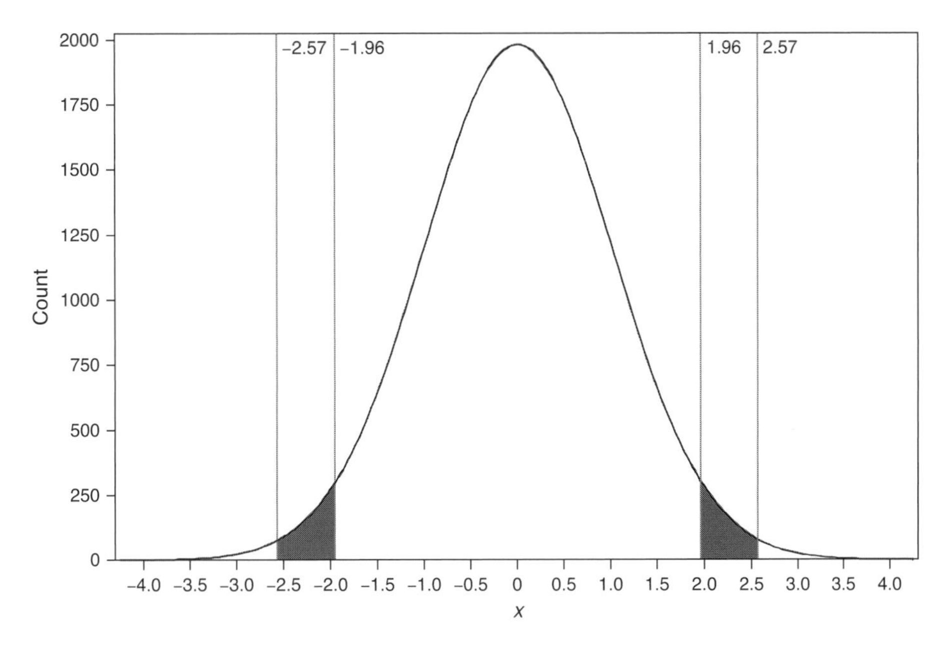

Figure 4.1 Areas of the normal curve delimited by −2.57, −1.96, 1.96, and 2.57.

compute two statistics and two parameters: the sample mean ($\bar{Y} = \frac{\sum Y}{n}$) and standard deviation ($s = \sqrt{s^2}$, where $s^2 = \frac{\sum (Y-\bar{Y})^2}{n-1}$) and the population mean ($\mu = \frac{\sum Y}{N}$) and standard deviation ($\sigma = \sqrt{\sigma^2}$, where $\sigma^2 = \frac{\sum (Y-\mu)^2}{N}$).

The reason we use the same formula for computing the sample and the population mean is that the former is an unbiased estimate of the latter. By this we mean that if we sample several sample means from a population (as we did in the previous chapter), most of them should approximate the true population mean. As we increase our sample size, we decrease the spread of sample means around the true population mean, or we reduce the standard error of the means $\sigma_{\bar{Y}} = \frac{\sigma}{\sqrt{n}}$. In contrast, the sample variance is a biased estimate of the population variance, and it must be adjusted by dividing the sums of squares by the degrees of freedom ($n - 1$) instead of n. When we are estimating population parameters it is frequently desirable to accompany an estimate by either a confidence limit or to state the confidence interval of our estimation. Both are defined and explained below.

4.2.1 Confidence limits and confidence interval

When we compute a sample mean with the purpose of estimating the population mean, it is desirable to accompany such an estimate with **confidence limits**; the interval between those limits is known as the **confidence interval**. The interval allows us to have a degree of confidence that we have covered the true value of the population sample within the interval. The degree of confidence in the limits is quite standard: we usually want to be 95% or 99% confident. If you look at Figure 4.1 you will see a standardized normal distribution with a mean of 0 and four z scores identified. The area between -1.96 and 1.96 covers 95% of the distribution while the area between -2.57 and 2.57 covers 99% of the distribution. Please confirm this with the normal table, keeping in mind that the table gives you the areas under the curve for one half of the distribution only. These z scores, shown in Figure 4.1, are used in the computation of the confidence limits, depending on which level of confidence we want. The formulae for the computation of the confidence limits *assuming knowledge of the parametric standard deviation* are shown below.

Formulae 4.1

Formulae for the confidence limits of a mean.

$$\textit{Lower limit} = \bar{Y} - (z_{0.99\ or\ 0.95)})(\sigma_{\bar{Y}})$$
$$\textit{Upper limit} = \bar{Y} + (z_{0.99\ or\ 0.95)})(\sigma_{\bar{Y}})$$

Where $\sigma_{\bar{Y}}$ is defined in formula 3.7.

As you well know, it is unlikely that we will ever know the true population standard deviation, which is required in formulae 4.1. Therefore, instead of using the standardized normal distribution, we use the ***t* distribution**, which is the distribution of ***t* scores**. The formulae for the computation of the confidence limits of a mean with the t distribution are shown in formulae 4.2. Just like a z score of 1.96 delimits 95% of the distribution (5% on each tail) there is a t score that delimits the 95% and 5% area of the t distribution curve for both tails. However, since the t distribution differs by sample size, there is a different t score associated with the 95%–5% cut-off point for each sample size. These cut-off points, better known as **critical values**, are arranged by the degrees of freedom $(n - 1)$ instead of by sample size (although computer programs will report results by sample size and by degrees of freedom). Table 4.1 shows selected values of the t distribution and Figure 4.2 shows the normal distribution with the t distribution superimposed. The larger the value of n, the more similar the t distribution is to the normal distribution. If $n = 30$ the t distribution is virtually the same as the normal one, and at $n = \infty$ the t distribution is identical to the normal distribution. For small samples the t distribution is flatter at the center, having more observations at the tails.

Formulae 4.2

Formulae for the confidence limits of a sample mean.

$$\textit{Lower limit} = \bar{Y} - (t_{0.99\ or\ 0.95})(s_{\bar{Y}})$$
$$\textit{Lower limit} = \bar{Y} + (t_{0.99\ or\ 0.95})(s_{\bar{Y}})$$

Where $s_{\bar{Y}} = \frac{s}{\sqrt{n}}$

Table 4.1 Selected critical values of the *t* distribution.

	Proportion in one tail			
	0.05	0.025	0.01	0.005
	Proportion in two tails			
df	0.10	0.05	0.02	0.01
1	6.314	12.706	31.821	63.657
5	2.015	2.571	3.365	4.032
9	1,833	2.262	2.821	3.250
10	1.812	2.228	2.764	3.169
13	1.771	2.160	2.650	3.012
15	1.753	2.131	2.602	2.947
20	1.725	2.086	2.528	2.845
25	1.708	2.060	2.485	2.787
30	1.697	2.042	2.457	2.750
40	1.684	2.021	2.423	2.704
60	1.671	2.000	2.390	2.660
120	1.658	1.980	2.358	2.617
∞	1.645	1.960	2.326	2.576

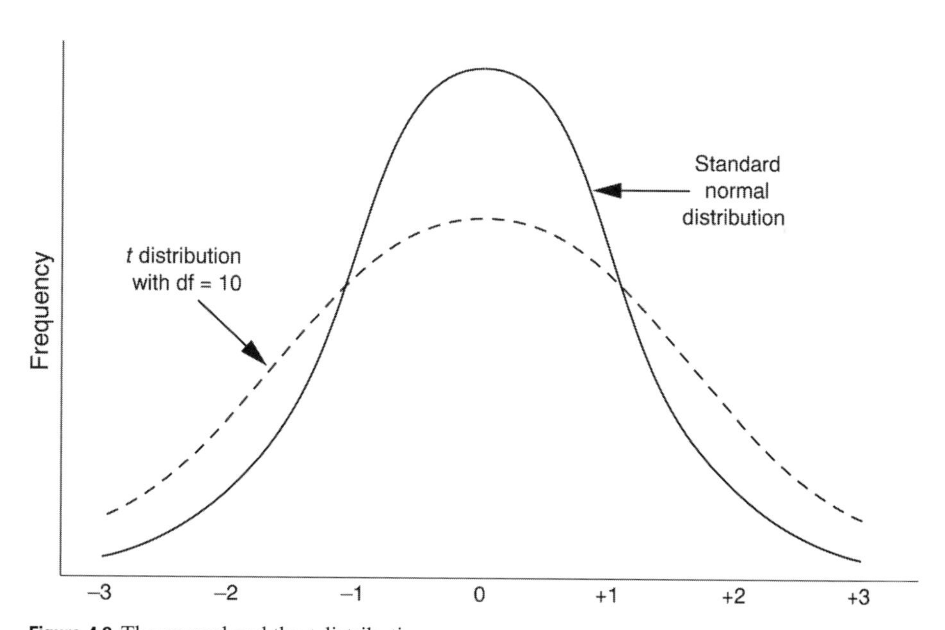

Figure 4.2 The normal and the *t* distributions.

Practice problem 4.1

An anthropologist is interested in the number of bulls owned by men in an agriculturalist community in a South west US town. He records a mean of $\mu = 4$ in a sample of 60 men. Compute the confidence limits for this mean with z scores using a population standard deviation of $\sigma = 2.5$ and with t scores using a sample standard deviation of $s = 2.75$. For both compute the confidence limits at 95% and 99% confidence limits.

Let us first use the standardized normal distribution, for which we need to compute $\sigma_{\bar{Y}} = \frac{2.5}{\sqrt{60}} = 0.3227$.

$$Lower\ limit_{0.95} = 4 - (1.96)(0.3227) = 3.367.$$
$$Upper\ limit_{0.95} = 4 + (1.96)(0.3227) = 4.632.$$

The confidence interval is $4.632 - 3.367 = 1.265$.

$$Lower\ limit_{0.99} = 4 - (2.57)(0.3227) = 3.171.$$
$$Upper\ limit_{0.99} = 4 + (2.57)(0.3227) = 4.829.$$

The confidence interval is $4.829 - 3.171 = 1.658$.

As expected the 99% confidence interval is broader than the 95% confidence interval. Let us now use the t distribution, for which we need to compute $S_{\bar{Y}} = \frac{2.75}{\sqrt{60}} = 0.3550$. You will notice that Table 4.1 does not have the critical values for $df = 60 - 1$. It is customary to err on the side of caution and to use a more conservative level, that is the next level below or $df = 40$. Therefore for the 95% confidence limits we use the critical value at $df = 40$ under the column 0.05 (proportion in two tails) which is 2.021. For the 99% confidence limits we use the critical value at $df = 40$ under the column 0.01 (proportion in two tails) which is 2.704.

$$Lower\ limit_{0.95} = 4 - (2.021)(0.3550) = 3.282.$$
$$Upper\ limit_{0.95} = 4 + (2.021)(0.3550) = 4.717.$$

The confidence interval is $4.717 - 3.282 = 1.435$.

$$Lower\ limit_{0.99} = 4 - (2.704)(0.3550) = 3.040.$$
$$Upper\ limit_{0.99} = 4 + (2.704)(0.3550) = 4.96.$$

The confidence interval is $4.96 - 3.04 = 1.92$.

The fact that the confidence intervals computed using the t distribution are broader than those computed using the standardized normal distribution is due precisely to the fact that we are using a different distribution. You can think of this as the price we have to pay for not knowing the population's parameters.

Let us compute again the confidence intervals with the t distribution, but let us use now a larger sample size of $n = 100$, therefore $S_{\bar{Y}} = \frac{2.75}{\sqrt{100}} = 0.2750$. The limits and interval are:

$$Lower\ limit_{0.95} = 4 - (2.021)(0.2750) = 3.444.$$
$$Upper\ limit_{0.95} = 4 + (2.021)(0.2750) = 4.555.$$

The confidence interval is 4.555 − 3.444 = 1.111.

$$Lower\ limit_{0.99} = 4 - (2.704)(0.2750) = 3.256.$$
$$Upper\ limit_{0.99} = 4 + (2.704)(0.2750) = 4.7436.$$

The confidence interval is 4.7436 − 3.256 = 1.4876.

As you can see, by increasing the sample size we reduced the size of our confidence interval. These results support our previous conclusion about the standard error of the mean: the error decreases if we increase our sample size.

4.2.2 Point estimation

When we compute a statistic with the purpose of estimating a population parameter we are engaging in point estimation. We have already noted that the sample $\bar{Y}$ is an unbiased estimate of the population μ. We know that even if the distribution of the population is not normal, the distribution of sample means will tend towards normality, and that the mean of the means will approximate the population parameter. We also know that as we increase sample size, error in our estimation (as measured by the standard error of the means) is reduced. The expected value of an unbiased estimate of a parameter is the true value of the parameter. If the parameter is denoted by θ and the estimator is denoted as $\hat{\theta}$ then the expected value of the estimator is $E(\hat{\theta}) = \theta$. I started this section with this discussion because the sample mean is the estimate of the population mean obtained by the two general methods of parameter estimation: least-squares estimation (LSE) and maximum likelihood estimation (MLE).

The least-squares method for estimating a parameter is easily graspable. If we have a data set consisting of the following observations: 7, 25, 58, and 99 (whose mean is 47.25), and if we compute the difference between each observation and the mean we obtain the following: $(7 - 47.25) + (25 - 47.25) + (58 - 47.25) + (99 - 47.25) = 0$. If, however, we square the difference between each observation and the mean we obtain: $(7 - 47.25)^2 + (25 - 47.25)^2 + (58 - 47.25)^2 + (99 - 47.25)^2 = 4908.75$. The least-squares estimate of the mean is the one that minimizes the sum of the squared differences between the observations and the mean. If our estimate of the mean had been 50, then the sum of the squares of the differences between the observations and the biased mean would have been 4939, which is higher than 4908.75.

I already mentioned the maximum likelihood approach as a powerful method for estimating parameters. But this is probably the last section in the book in which I will discuss it because all statistical techniques discussed in the rest of the book will rely on a least-squares approach. Briefly, if we observe a random variable with sample size n whose parameters are $\theta_1, \theta_2, \ldots \theta_k$, then the maximum likelihood estimator of the k parameters are the values which maximize the probability of sampling the observed data. In other words, the maximum likelihood estimate is one which produces the probability distribution that makes the observed data most likely (Myung 2003). The maximum likelihood estimate of a sample mean is the one that maximizes the likelihood function of the observed data. Maximum likelihood estimation requires knowledge of calculus, which is not assumed in the reader of this book. In contrast, the least-squares method for estimation relies on algebra, which is assumed.

4.3 Hypothesis testing

In chapter 1 we discussed the use of statistical analysis in the scientific endeavor. Not only do statistics allow us to summarize and understand our data, but they permit us to test hypotheses about the population from which data are collected. Fisher (1993) himself mentions that the people who developed (now classic) statistical tests did so in the midst of a research problem with the purpose of answering a scientific question. This section introduces the reader to principles of hypothesis testing, and how statistics are used for this scientific purpose. Examples of such tests are included. As I discussed before, what I am presenting here is the classical approach to hypothesis testing.

4.3.1 The principles of hypothesis testing

Earlier in the book it was mentioned that a scientific hypothesis is one which can be tested. This testing proceeds (basically) in the following manner:

1. The null (H0) and alternative (H1) hypotheses are stated.
2. The level of statistical significance is established.
3. The sample is collected.
4. The sample is compared with the null hypothesis' parameters, and a conclusion is reached about which hypothesis to accept. Each of these points is discussed below.

(1) The null (H0) and alternative (H1) hypotheses are stated

The first step in hypothesis testing is for the researcher to propose a null hypothesis about the population under study, specifically about the population's parameters. An example of such a **null hypothesis** could be: "In this population, the mean age at marriage is $\mu = 20$ years." The H0 is called the null hypothesis because it proposes that there is no difference between the true value of the population's parameter and that stated in the H0 and because it states that the population parameter has not been affected by a treatment effect (the treatment has had a null effect on the population).

The **alternative or scientific hypothesis** (H1) proposes that the treatment has had an effect, and that the true parameter of the population from which our sample was obtained is different from that proposed by the null hypothesis. For example, a researcher may know that the mean age at marriage in a country is $\mu = 20$ years. However, he is interested in studying the age at marriage in a community which is ethnically distinct from the larger national culture. The anthropologist suspects that cultural practices such as age at marriage differ in this small community from those of the larger national culture. The researcher then proposes the following hypotheses: H0: $\mu = 20$ years. H1: $\mu \neq 20$ years. The alternative hypothesis here proposes that the treatment effect (ethnicity) has a significant effect on the population, making the small community's μ *statistically significantly* different from 20.

It is interesting that even though H1 is the hypothesis generated by the research interest, H0 is the one being tested. If H0 is not rejected, it is accepted as the best explanation of the facts, but it has not been proven. Indeed, some statisticians feel that it is more proper to say "fail to reject" than to say "accept" the null hypothesis. Fisher himself said that significance testing is not a means for accepting a hypothesis, as he was focusing on hypothesis rejection as opposed to acceptance (Fisher 1993). If H0 is rejected, H1 is accepted as the best explanation of the facts, but it has not been proven either. Indeed, this is how scientific knowledge progresses: researchers accept the most likely explanation of the facts, while being totally open about the possibility that such explanation may be falsified in the future.

A very important topic in hypothesis testing and in science in general is discussed by S. J. Gould in his essay titled "Cordelia's dilemma" (1993). The issue concerns the (very human) desire to "prove our hypothesis" (although we never prove anything in science). This is a very understandable desire: if a researcher spends much time and effort researching a topic, thinking that (H1) X affects Y, only to find out that it does not, she will surely be disappointed. But what we all need to realize is that negative results (H0 is not rejected) are as important as positive ones (H0 is rejected). Both kinds of results inform us about the behavior of our statistical population. Bramblet (1994) discusses that in primatology, behavioral/biological significance may be divorced from statistical significance. If one rare event marks the change of an individual's status in a group's hierarchy, we still need the many hours of observation in which this occurrence did not happen for us to understand how significant this rare event was to the group. The same can be said of paleoanthropological studies: even if we are interested in the brief, episodic, and significant periods of directional change in the fossil record, we still need to document and understand the long periods of no directional change (Gould 1993).

(2) Establish the significance level

Before the researcher collects a sample to test his hypothesis, he must decide how different a sample mean $\bar{Y}$ must be from the population μ proposed by H0 for the H0 to be rejected. The investigator faces the following problem: if he is to sample repeatedly from a population, the mean of different samples will differ. Indeed, in the previous chapter we performed an "experiment" in which we obtained all possible samples of size $n = 2$ from a known population of size $N = 3$ and whose $\mu = 20$. We *know* that every single one of these samples was collected from the population. However, whereas some sample means were equal to the population mean, others were rather different. Therefore, the problem in hypothesis testing is to establish how deviant a sample mean must be from the μ proposed by the H0 for the researcher to decide to reject the H0. The solution to this problem is the following: we compute a z score for the $\bar{Y}$ *using the population information proposed by the H0*, and see where in the population such a z score falls. It is a widely accepted statistical convention that if a z score falls within

95% of a population's distribution around the μ, it is considered not to differ in a statistically significant manner from the population's mean. The remaining 5% of the distribution, equally divided between both tails of the distribution (2.5% on each side), is considered to be significantly different from the population's mean. Accordingly, 95% of the distribution is called **the acceptance region** (if a z score falls within it, the H0 is accepted), and 5% of the distribution, equally divided between both tails is called **the rejection or critical region**. The rejection area is referred to as alpha (α), and its value is established by the researcher to be (usually) 5% or 0.05. The acceptance region is obtained by subtraction: $1 - \alpha$.

We will initially use the table of areas under the normal curve to test hypotheses: what we need to find is a z score whose C column value is 0.025, since we are dividing equally 0.05, the α level at which we will reject H0 between both tails. Such a z score is 1.96, whose column C's value is 0.025. Therefore, if a z score is equal to or greater than 1.96, it will be considered to be significantly different from the population's mean (see Figure 4.1 and Table 3.5).

(3) Data collection

After the H0 and H1 have been proposed, and the researcher has chosen a particular α level, the sample may be collected. As mentioned previously, a discussion of research design and data collection is outside the scope of this book.

(4) Compare the sample with the null hypothesis, and reach a conclusion about the latter

At this stage, the researcher computes the sample mean's z score as done in the last chapter ($z = \frac{\bar{Y}-\mu}{\sigma_{\bar{Y}}}$). If the z score is ≥ 1.96 then it will be considered to be significantly different from the μ proposed by the H0.

In our fictitious research project on age at marriage in a small ethnically distinct community, we know from census data that for the entire country $\mu = 20$ and $\sigma = 5$ years, and we propose the following hypotheses: H0: $\mu = 20$ years. H1: $\mu \neq 20$ years. What would our decision be if we collected a sample of $n = 30$ whose $\bar{Y} = 19$? What would our decision be if we collected a sample of $n = 30$ whose $\bar{Y} = 18$?

$\bar{Y}$	n	$\sigma_{\bar{Y}}$	z	Decision
19	30	$\frac{5}{\sqrt{30}} = 0.91$	$\frac{19-20}{0.91} = -1.1$	Accept H0
18	30	$\frac{5}{\sqrt{30}} = 0.91$	$\frac{18-20}{0.91} = -2.2$	Reject H0

If the sample mean had been 19, we would decide that the mean of the population from which we obtained our sample was indeed $\mu = 20$. Instead, a sample mean of 18 would be considered to be too different from $\mu = 20$ (its z score was > than 1.96), and would be declared not to have been sampled from the population proposed by H0.

Practice problem 4.2

For this exercise, our population's parameters are $\mu = 20$ and $\sigma = 5.77$. The population size is N = 3 with $y_1 = 10, y_2 = 20$, and $y_3 = 30$. Let us obtain five different samples of $n = 2$ whose respective means are 10, 15, 20, 25, and 30. Let us compute for each sample mean a z score to test the null hypothesis that the population from which these samples were obtained has a mean of 20. Thus: H0: $\mu = 20$. H1: $\mu \neq 20$. If the z score is equal to or greater than 1.96, we reject H0.

$\bar{Y}$	n	$\sigma_{\bar{Y}}$	z	Decision
10	2	$\frac{5.77}{\sqrt{2}} = 4.1$	$\frac{10-20}{4.1} = -2.44$	Reject H0
15	2	$\frac{5.77}{\sqrt{2}} = 4.1$	$\frac{15-20}{4.1} = -1.22$	Accept H0
20	2	$\frac{5.77}{\sqrt{2}} = 4.1$	$\frac{20-20}{4.1} = 0.0$	Accept H0
25	2	$\frac{5.77}{\sqrt{2}} = 4.1$	$\frac{25-20}{4.1} = 1.22$	Accept H0
30	2	$\frac{5.77}{\sqrt{2}} = 4.1$	$\frac{30-20}{4.1} = 2.44$	Reject H0

This example is of pedagogical use: it illustrates the relationship between hypothesis testing and probability theory. All other hypothesis tests presented in this book rest on the same principles: if the outcome of an experiment or survey is statistically unlikely to have been obtained from a population, then it is said not to have come from such a population. Unlikely here is defined as falling in the rejection area. Clearly, however, our decision to reject the null hypothesis in this exercise was in error: we *know* that the samples with $\bar{Y} = 10$ and $\bar{Y} = 30$ did come from the population. The next section discusses errors in hypothesis testing.

4.3.2 Errors and power in hypothesis testing

When testing a hypothesis, a researcher may commit either of two errors. (1) Rejecting the null hypothesis when in fact it is true (type I error; the true mean age at marriage in our population is not different from 20 but we rejected the null hypothesis saying that $\mu = 20$) and (2) Accepting the null hypothesis even though it should be rejected (type II error; the true mean age at marriage in our population is not equal to 20, but we did not reject the H0 saying that $\mu = 20$). We discuss each error below.

	Null hypothesis is true	Null hypothesis is false
Decision: reject H0	Type I error	Correct decision
Decision: accept H0	Correct decision	Type II error

4.3.2.1 Type I error (α)

A type I error was illustrated in the previous section: the null hypothesis was correct even with the samples whose means were $\bar{Y} = 10$ and $\bar{Y} = 30$, yet we decided to reject H0 because these outcomes fell outside the acceptance region. How likely are we to commit this error? The answer is that the probability is equal to α, in this case 0.05. Therefore, when we decide to reject a null hypothesis, we also state that we know that we could be making an error in our decision, and we actually state how likely we are to be making such an error: the probability is α. This means that if $\alpha = 0.05$, then 1 out of 20 trials ($1/20 = 0.05$) or samples obtained from the population could lead the researcher to erroneously reject H0. If the reader thinks this is too high a risk, then she could decide to decrease her rejection region to, say, 0.01. This would mean that now 1 out of 100 trials could lead to a type I error. The reader could go even further and use an alpha level of 0.001, in which case only 1 out of 1000 trials would result in a type I error. The *z* scores associated with these α levels are:

Alpha	z score	Column C
0.05	1.96	0.025
0.01	2.58	0.0049 = 0.005
0.001	3.3	0.0005

However, the reader has probably grasped by now that this exercise, while decreasing α, has only increased the probability of committing a type II error (accepting H0 when in fact it is false). This is why an α level of 0.05 is considered a good middle ground.

4.3.2.2 Type II error (β)

The topic of what exactly the probability associated with type II error (or β) is, is frequently left out of statistics textbooks. This is so because in most cases such probability cannot be computed since it requires the H1 to be stated as precisely as the H0 is. For example, in our age-at-marriage project, we would need to state H0: $\mu = 20$ and (e.g.) H1: $\mu = 25$. As the reader can see, this level of precision requires knowledge about two populations, knowledge which is usually unavailable. However, the probability of committing a type II error is illustrated here with another purely pedagogical example.

Let us assume that we have total knowledge of two populations (H0 and H1), both of which have size N = 3. The first population's observations are 10, 20, and 30 (with a mean of $\mu = 20$). The second population's observations are 15, 25, and 35 (with a mean of $\mu = 25$). If we sample with replacement all possible samples of size

$n = 2$ we obtain the following frequency distributions for the sample means of both populations:

Population 1: H0			Population 2: H1		
Sample means	Frequency	Percent	Sample means	Frequency	Percent
10	1	$1/9 = 0.111$	15	1	$1/9 = 0.111$
15	2	$2/9 = 0.222$	20	2	$2/9 = 0.222$
20	3	$3/9 = 0.333$	25	3	$3/9 = 0.333$
25	2	$2/9 = 0.222$	30	2	$2/9 = 0.222$
30	1	$1/0 = 0.111$	35	1	$1/0 = 0.111$
$\mu = 20$	$\Sigma f = 9$	$0.999 \approx 1$	$\mu = 25$	$\Sigma f = 9$	$0.999 \approx 1$

The *probability of committing a type II error* is the proportion of H1's population which overlaps with the acceptance region of H0's distribution. The acceptance region of H0's distribution included only samples whose means are 15, 20, and 25 (because the H0 was not rejected for these means), and all of these overlap with H1's distribution. If the probability of those sample means is added, we obtain the probability of committing a type II error. Thus, $\beta = 0.222 + 0.333 + 0.222 = 0.777$.

4.3.2.3 Power of statistical tests $(1 - \beta)$

The previous section brought us to the topic of **power** in a statistical test, which is the ability of a test to reject H0 when H0 is false. Since a type II error (β) is the acceptance of H0 when in fact it should be rejected, power is defined as $1 - \beta$. Therefore, in the example presented in the previous section, the power of our test was $1 - 0.777 = 0.223$. Statistical power, just like the probability of a type II error, depends on how different the means of the populations proposed by H0 and H1 are. If H0: $\mu = 20$ and H1: $\mu = 20$, then the overlap of H1's distribution with the acceptance region of H0 is total. In this case, β would be equal to the entire acceptance region, which (if $\alpha = 0.05$) would be $1 - 0.05 = 0.95$ and power would be $1 - 0.95 = 0.05$. In sum, the more different the two parametric means are, the higher the power of the test.

For the computation of β and $1 - \beta$ we usually rely on the normal distribution table to establish the proportion of H0's population acceptance region which overlaps with H1's distribution (instead of relying on a frequency distribution of sample means as shown above). We proceed as follows:

1. Establish the 95% cut-off points of H0's distribution.
2. Compute a z score by subtracting from such cut-off point the value of H1's μ, and dividing by the standard error of the means.
3. Column C of Table 3.5 tells us the proportion of the distribution beyond the z score. This is the proportion of H0's acceptance region which overlaps with H1's distribution. This procedure is illustrated below.

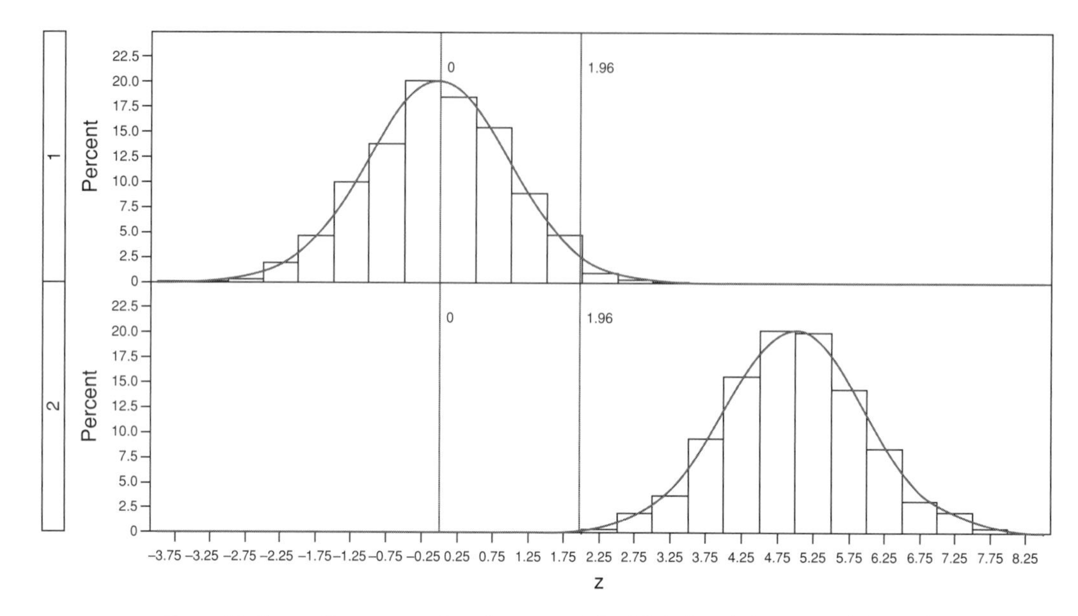

Figure 4.3 Computing power when the mean of the H0 is 0 and the mean of the H1 is 5.

Figure 4.3 shows the distribution of both H0 ($\mu = 0$, $\sigma = 1$) and H1 ($\mu = 5$, $\sigma = 1$). The right-hand limit to the acceptance region of H0 is marked. The question we ask is what proportion of H0's acceptance region overlaps with the distribution of H1. We will test the hypothesis with a sample size of $n = 10$ (since $\sigma = 1$, then $\sigma_{\bar{Y}} = \frac{1}{\sqrt{10}} = 0.32$).

We follow these steps: we compute a z score by subtracting 5 (H1's μ) from 1.96, the cut-off point of H0's acceptance region, and divide by the standard error of the means (0.32). Then we look for the area beyond this z score (column C). This is the area of the acceptance region of H0 which overlaps with H1's distribution. Thus: $z = \frac{5-1.96}{0.32} = 9.5$. The area beyond this z score is ≈ 0. The probability of committing a type II error approaches 0 and the power of this test is ≈ 1. Power can never be equal to 1, and the probability of committing a type II error can never be equal to 0. This is so because distributions of populations approach asymptotically the horizontal axis, and extend infinitely.

Practice problem 4.3

Compute β and $1 - \beta$ in the following examples. We will keep the same H0 ($\mu = 0$, $\sigma = 1$), but alter H1. In all cases, our sample will have a size of $n = 20$, so $\sigma_{\bar{Y}} = \frac{1}{\sqrt{20}} = 0.22$.

1. H0: $\mu = 0, \sigma = 1$, H1: $\mu = 3, \sigma = 1, n = 20, \sigma_{\bar{Y}} = 0.22$.

   In this case $z = \frac{3-1.96}{0.22} = 4.73$. The area beyond this z score is ≈ 0. This is the probability of committing a type II error, and, therefore, the power of this test is ≈ 1 (Figure 4.4).

2. H0: $\mu = 0, \sigma = 1$, H1: $\mu = 2, \sigma = 1, n = 20, \sigma_{\bar{Y}} = 0.22$.

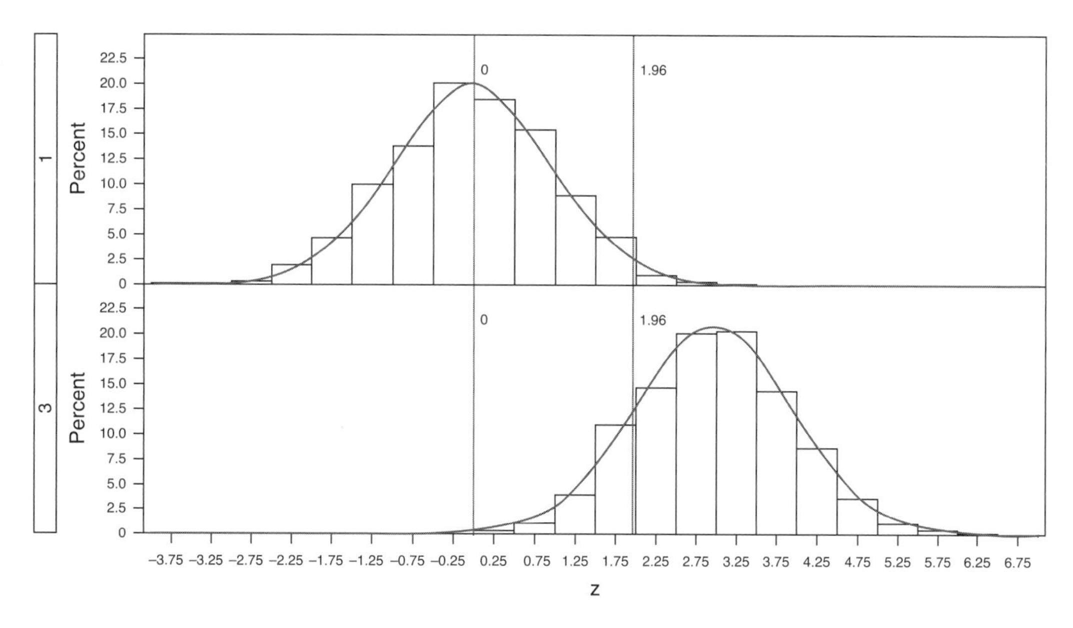

Figure 4.4 Computing power when the mean of the H0 is 0 and the mean of the H1 is 3.

For this example $z = \frac{2-1.96}{0.22} = 0.18$. The area beyond a z score of 0.18 is 0.4286. This is the probability of committing a type II error, and, therefore, the power of this test is $1 - 0.4286 = 0.5714$. The less different the two populations, the less powerful the test (Figure 4.5).

An interesting aspect of statistical power is that it can be increased by increasing the sample size. Thus, if instead of obtaining a sample of size $n = 20$ in the examples above, we obtained one of $n = 40$, we would have increased our power while decreasing our probability of a type II error. Thus: $\sigma_{\bar{Y}} = \frac{1}{\sqrt{40}} = 0.16$. In the first example, we had: 1. H0: $\mu = 0, \sigma = 1$, H1: $\mu = 3, \sigma = 1$. Thus $z = \frac{3-1.96}{0.16} = 6.5$. The area beyond this z score is even closer to 0 than that of the z score computed with $\sigma_{\bar{Y}} = 0.22$. Thus, the power of this test is very near 1.

In the second example we had: 2. H0: $\mu = 0, \sigma = 1$, H1: $\mu = 2, \sigma = 1$. Thus $z = \frac{2-1.96}{0.16} = 0.25$. The area beyond this z score is 0.4013, a smaller probability of committing a type II error than that computed with a sample size of $n = 20$. The power of the test is now $1 - 0.4013 = 0.5987$ instead of 0.5714.

Before commencing a research project (for example, at the grant-writing stage) it is sometimes desirable to conduct a **power analysis** to estimate the necessary sample size to demonstrate the presence of a treatment effect (reject the null hypothesis when in fact it is false). Including a power analysis in a grant proposal would certainly indicate that you are not pulling out of a hat your target sample size, but that you have valid reasons to wish to collect such a sample. If you do a power analysis you specify the desired α level, the desired power level and your best guess of the magnitude of the treatment effect.

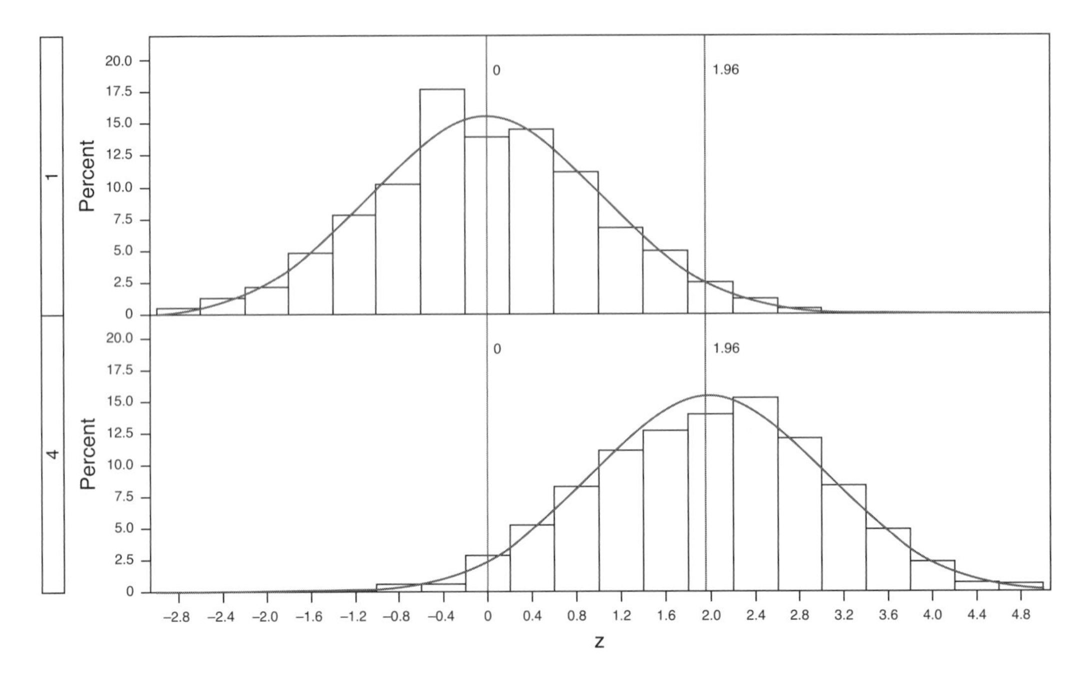

Figure 4.5 Computing power when the mean of the H0 is 0 and the mean of the H1 is 2.

A large treatment effect would produce a very different μ in the H1 when compared with the μ in the H0 while a small one would produce very similar population means. Cohen (1992) has come to the rescue of researchers in all fields with his "A power primer," where he provides the target sample size for the most commonly used statistical tests with different degrees of freedom or number of groups, with α levels of 0.01, 0.05, and 0.1, for small, medium, and large treatment effects at power $= 0.80$. Although there are computer programs available for obtaining this information, I dare say that you might find everything you need in the paper by Cohen (1992) or in his larger *Statistical Power Analysis for the Behavioral Sciences* (Cohen 1988).

As we mentioned above, decreasing the probability of committing a type I error by decreasing the α level only results in an increasing probability of committing a type II error. The solution to obtaining acceptable levels of both types of errors is to increase sample size to the desirable numbers computed by Cohen or by a computer package. The advantage of doing a power analysis is that we do not unnecessarily sample too large a sample which does not serve a statistical purpose. Too large a sample is costly in terms of time and money. The reader is strongly encouraged to compute the target sample size when planning a new research project.

4.3.3 Hypothesis tests using *z* scores

As was mentioned previously, although the null hypothesis is tested, the alternative hypothesis is the one usually generated by the scientist's research interest. Thus, whereas the researcher may want to show that there is a treatment affecting the behavior of his sample, the null hypothesis will propose that there is none. If H0 is rejected, then H1 is accepted as the most likely explanation: the data support the proposal that there is a treatment effect. The following examples are of pedagogical value only, and do not reflect actual data. Their purpose is to illustrate hypothesis testing with z scores.

1. An anthropologist is interested in studying the effect of religious affiliation on fertility. She works in a community which consists of a distinct religious group, and suspects that in this community large families are valued more than they are in the wider national community. The researcher wishes to determine if the small community has a significantly different mean family size from that of the national society. The anthropologist collects national census data and finds out that for the entire country the mean number of children per household is $\mu = 4$ with $\sigma = 1.8$. Thus H0: $\mu = 4$ and H1: $\mu \neq 4$. The anthropologist decides to use a 0.05 level for α. The investigator obtains a sample of size $n = 10$ families and computes a sample mean of $\bar{Y} = 6$. Let us test the null hypothesis with a z score. Thus: $\sigma_{\bar{Y}} = \frac{1.8}{\sqrt{10}} = 0.57$ and $z = \frac{6-4}{0.57} = 3.5$. The probability of finding this or a more extreme value is given by column C, and it is 0.0002. Since the probability is less than 0.05 (or, since the z score is larger than 1.96, the 0.05 cut-off point in the normal distribution table), the null hypothesis is rejected. The researcher also knows that she could be making a type I error, but the probability of that is rather low: it is 0.0002.
2. A researcher is interested in studying if a particular horticultural group consumes different numbers of animal species seasonally. To this effect, the anthropologist divides the year into the wet and dry season, and uses the group's diet of the wet season to generate the null hypothesis. The investigator determines that the mean number of animal species consumed during the wet season is $\mu = 10$ with a standard deviation of $\sigma = 3$. He returns to the community during the dry season and obtains samples from 20 households, and finds out that the mean number of animal species consumed during the dry season is $\bar{Y} = 9$. Therefore: H0: $\mu = 10$. H1: $\mu \neq 10$. The null hypothesis will be rejected at the usual $\alpha = 0.05$. Thus: $\sigma_{\bar{Y}} = \frac{3}{\sqrt{20}} = 0.67$ and $z = \frac{10-9}{0.67} = 1.49$. In this example the researcher fails to reject the null hypothesis, and accepts that the number of animal species consumed is stable through the year.
3. A team of anthropologists is interested in establishing if in the community of their study males and females have a different number of words to refer to soil quality (for example, the mineral content). It is known that for the male population, the mean number of words for soil quality is $\mu = 4$ with a standard deviation of $\sigma = 1.1$. This information will be used as the null hypothesis' population. A female anthropologist takes a sample of $n = 30$ females and determines that the mean number of words for soil quality is $\bar{Y} = 7$. Therefore: H0: $\mu = 4, \sigma = 1.1$. H1: $\mu \neq 4$. The null hypothesis will be rejected at the usual $\alpha = 0.05$. Thus: $\sigma_{\bar{Y}} = \frac{1.1}{\sqrt{30}} = 0.20$ and $z = \frac{7-4}{0.20} = 15$. This is a highly significant result, which leads to the rejection of the H0 with a high confidence that the decision is not in error. The probability that a type I error was committed is close to 0.

4.3.4 One- and two-tailed hypothesis tests

The reader has probably noticed that up to this point, whereas all our null hypotheses have been stated exactly (say H0: $\mu = 20$) our alternative hypotheses have not (H1: $\mu \neq 20$) (except for the case when we illustrated the computation of β and $1 - \beta$). In two-tailed or two-way hypothesis testing, the researcher does not indicate in what direction the treatment could affect the data. Thus, the researcher does not say that he suspects (for example) that members of a certain religious group are likely to have larger or smaller family sizes. Two-tailed tests are customary and widely accepted. Indeed, researchers rarely mention in writing that a test was two-tailed.

However, there are some situations which call for a one-tailed approach. **One-tailed tests are directional**, that is, they state in what direction the treatment is expected to affect the data. Notice that both the H0 and the H1 are changed from what we did previously:

Two-tailed	One-tailed
	For a positive treatment effect:
H0: $\mu = 4$ and H1: $\mu \neq 4$	H0: $\mu \leq 4$ and H1: $\mu > 4$
	For a negative treatment effect:
H0: $\mu = 4$ and H1: $\mu \neq 4$	H0: $\mu \geq 4$ and H1: $\mu < 4$

Please notice that whereas the alternative hypothesis states that the mean will be greater or lesser than H0's μ, the null hypothesis states that the mean could be affected in an opposite manner from that stated by H1, or that it could be not affected at all. In other words, there are three possibilities which need to be covered in a one-tailed test: the direction stated by H1, the opposite direction, and no change in direction. The last two possibilities are stated in the H0.

One-tailed tests may be used if the researcher either knows that the treatment cannot affect the data in any but the direction stated by H1, or if he is simply not interested in any other direction. Conceivably, if a drug is tested in patients, the researchers may truly know that the drug will only improve the patient's health.

The main reason one-tailed tests are not frequently performed (so that researchers disclose in publications that they applied a one- instead of a two-tailed test) is that such tests are associated with a greater probability of a type I error, even if the same α level is used. If the investigator is going to reject H0 by looking at one tail only and wishes to use $\alpha = 0.05$, then she needs a cut-off point of H0's distribution beyond which all of the 5% of the distribution is found. This cut-off point is 1.645, which is actually not found in our table, so we use 1.65 instead. Previously, we rejected the H0 if the z score was greater than or equal to 1.96, which has 0.025 of the distribution beyond it (because we were considering 2.5% on each of two tails). In a one-tailed test, however, we can reject the null hypothesis on one side only, so we use a cut-off point with 0.05 beyond it.

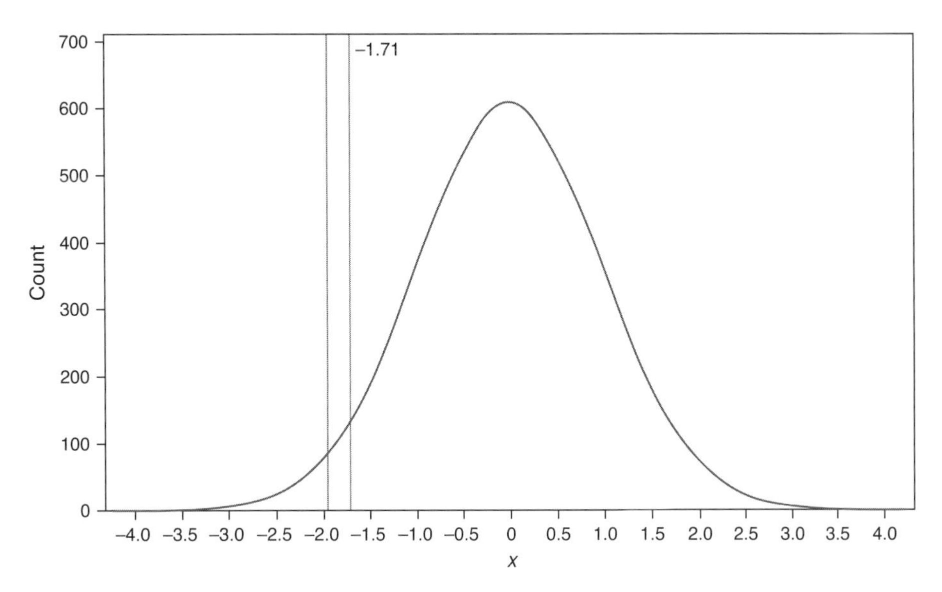

Figure 4.6 The location of z score of -1.71, which is significant for a one-tailed test but would not have been significant for a two-tailed test. The second line shows the position of $z = -1.96$.

The difference between these two approaches is best illustrated with an example. Let us say that an anthropologist has been working in a country in which the national mean number of children produced by women at the end of their reproductive career is $\mu = 6$ with a $\sigma = 2$. The researcher is interested in determining if women who have had more years of higher education have lower fertility. The researcher chooses to use a one-tailed test because he feels there is strong enough evidence to suggest that higher education lowers, but never increases, the fertility of women. The anthropologist then takes a sample of size $n = 32$ higher-educated females and determines that the mean number of children produced by these women at the end of their reproductive careers is $\bar{Y} = 5.4$. The researcher decides to use an $\alpha = 0.05$, so the cut-off point will be 1.65 instead of 1.96. The hypotheses are stated as follows: H0: $\mu \geq 6$ and H1: $\mu < 6$. A z score is computed as done previously: $\sigma_{\bar{Y}} = \frac{2}{\sqrt{32}} = 0.35$. $z = \frac{5.4-6}{0.35} = -1.71$. Since the z score is greater in absolute value than 1.65, the null hypothesis is rejected. If the researcher had decided to use a two-tailed test he would not have rejected H0 because the z score is less than 1.96 (see Figure 4.6).

4.3.5 Assumptions of statistical tests

When a hypothesis is to be tested with a statistical test, the test requires a number of assumptions from the data and the manner in which the data were collected. Some of these assumptions are specific to the particular test, so they will be discussed in the appropriate chapter. Some, however, are shared by many tests, so they are discussed here.

1. *Random sampling*. Most tests assume that the sample was collected giving each member of the population an equal chance of being selected. Only with a random sample do we have an accurate representation of all members of the population.
2. *Independence of observations*. This is an assumption which is met by a sound sampling design. The assumption means that if an observation is collected, it will not influence which observation is collected next. Thus, the variates are independent from each other. Anthropologists who work with historical demographic data may confront a source of data which does not meet this assumption. If the researcher collects the number of deaths per month in a community, and the community is affected by a cyclical mortality pattern, then the number of deaths in a month is not independent from the number of deaths in the next month. For example, the datum of a month in the rainy season will have a high correlation with the data of the other months of the rainy season. In this case, statistical analyses that are designed to deal with auto-correlated data must be applied (Madrigal and Koertvelyessy 2003).
3. *The data are normal*. As was mentioned previously, it is known that many continuous numerical variables *are* normally distributed. Thus, it is known that the principles about hypothesis testing which apply to the normal distribution also apply to (e.g.) height and many other variables. In other words, we can safely say that 68.27% of all subjects are found within the first positive and the first negative standard deviations, that 95.45% are found within the first 2 positive and negative standard deviations, and so forth. Therefore, we can also use the principles of hypothesis testing already demonstrated in this chapter, and decide that if an outcome is found within 95% of the distribution it will not be considered to be *statistically significantly* different from the mean. If the outcome is found in the 5% rejection area, then we can decide that it is statistically significantly different from the mean. The point here is that if a population is normally distributed, we can use these principles to test hypotheses about the population.

What if the population is not normally distributed? Then we cannot use statistical tests that assume normality. If we did, we could be rejecting the null hypothesis in error, not because of the type I and type II errors implicit in every statistical test, but because the population does not follow a normal distribution. The assumption of normality is frequently not given much importance because most samples of size $n = 30$ *are* normally distributed. However, PASW and SAS offer extremely easy ways of testing if data are normally distributed so that there is no reason not to test if our data are normally distributed.

If a data set is found to be non-normal, then the researcher has two options open. (1) A non-parametric test is applied. Non-parametric tests have few assumptions, and are easy to compute in small data sets. The negative aspect about non-parametric tests is that they are less powerful (statistically speaking) than are parametric ones. This book covers non-parametric statistical tests in chapter seven. (2) The data set can be transformed, so that it does become normal. Fortunately, this is a fairly easy task with most statistical computer packages. The purpose of transforming the data is to make the data conform to a normal distribution and to be able to test a hypothesis with a particular statistical test which assumes normality. The outcome of the test is not under the control of the researcher. Transformations are discussed in chapter ten.

4.3.6 Hypothesis testing with the *t* distribution

At this point it is worthwhile to review what we have covered so far in terms of hypothesis testing. We have tested a null hypothesis that proposes that a sample mean $\bar{Y}$ belongs to a population whose parametric mean is μ. The null hypothesis was tested by computing a z score as: $z = \frac{\bar{Y}-\mu}{\sigma_{\bar{Y}}}$ where $\sigma_{\bar{Y}} = \frac{\sigma}{\sqrt{n}}$. Notice that the hypothetical value of μ *is given by the researcher*. For example, we proposed that if a horticultural group did not experience seasonal variation in the number of animal species it consumed during the entire year, then the sample means obtained during different times of the year were the same, because they sampled the same population and thus had the same μ. Thus, it is proposed that $\bar{Y}_{\text{wet season}} = \bar{Y}_{\text{dry season}} = \mu$. We also tested the hypothesis that in a particular society, the number of words for soil quality used by males and females was the same, or that the parametric number of words (μ) for females was the same as that for males. Thus, a researcher can always propose a hypothetical value for a μ, and test if the obtained sample differs from it. But in all these examples we assumed knowledge of σ, knowledge we do not usually have. Therefore, we need another means to test hypotheses "in the real world," that is, without knowledge of the population's σ. Our needs are met by computing t scores instead of z scores, and by using the t distribution, instead of the normal distribution.

Just like we can transform any sample mean into a z score, we can transform any sample mean into a t score. The difference is that we use the sample formula for computing the standard error of the means. The formula for the t score is shown in formula 4.3.

Formula 4.3

A sample mean's *t* score

$$t = \frac{\bar{Y} - \mu}{s_{\bar{Y}}}, \text{ where}$$

$$s_{\bar{Y}} = \frac{s}{\sqrt{n}}$$

The main difference between a t score and a z score is that the former uses the sample standard deviation instead of the population standard deviation for the computation of the standard error of the mean (see formula 4.3).

The t distribution is used for hypothesis testing in the same manner as is the normal distribution. That is, we compute a t score and ask what the probability is of obtaining this or a more extreme score. We will continue to use the same alpha level, that is, if the probability associated with a particular score is less than or equal to 0.05, we will declare the score's difference from μ statistically significant. Because the shape of the t distribution is symmetrical, it does not matter if a t score is positive or negative. The

t distribution also asymptotically approaches the horizontal axis, so that we can never say that the probability associated with a very unusual score is 0, we can only say that such probability approaches 0.

Since there is a t distribution for each df from 1 to ∞, the cut-off point or **critical value** (CV) between the distribution's acceptance and rejection areas is specific to each df's distribution. Therefore, whereas with the normal distribution, we knew that a z score whose value was 1.96 had a two-tailed 0.05 probability of occurrence, and we used it as a cut-off point, there is a different cut-off point for each t distribution. Thus, if we wish to determine if a t score is significantly different from the parametric mean, we need to look at the table, having established the significance level and the degrees of freedom. In case a one-tailed hypothesis is being proposed, we need to use a critical value which has all of the distribution's 5% on its side, as opposed to having 2.5% equally distributed on each tail. However, two-tailed hypotheses are usually preferred (a smaller t value would be considered to be significant in a one-tailed test as opposed to a two-tailed test).

Let us review the steps involved in hypothesis testing using the t distribution:

1. The null (H0) and alternative (H1) hypothesis are stated. The null hypothesis will usually be of the sort H0: μ = a hypothetical number.
2. The level of statistical significance is established.
3. The sample is collected and the t score is computed.
4. The sample is compared with the null hypothesis' parameters, and a conclusion is reached about which hypothesis to accept.

4.3.7 Hypothesis tests using t scores

We use here the same examples we used in section 4.3.3, where we tested these hypotheses with z scores. However, since we use here the sample formula to estimate s instead of the population formula to estimate σ, the value of $s_{\bar{Y}}$ and $\sigma_{\bar{Y}}$ are different.

1. In this example, the researcher is interested in investigating whether a religious isolate has a significantly different mean number of children per household from that of the national society. The investigator obtains a sample of size $n = 10$ families, whose $\bar{Y} = 6$ and $s = 2$. The researcher uses census information to propose the null hypothesis' μ. The H0 is: $\mu = 4$. The H1: $\mu \neq 4$; thus, the test is two-tailed. An α level of 0.05 will be used. The t score is computed as follows: $s_{\bar{Y}} = \frac{2}{\sqrt{10}} = 0.63$. $t = \frac{6-4}{0.63} = 3.17$. At $\alpha = 0.05$ for df $= 10 - 1 = 9$, the critical value is 2.262 (see Table 4.1). Thus, we reject H0 at a 0.05 level. As a matter of fact, our t statistic is greater than the critical value at 0.02% (CV = 2.821), giving us more confidence that we are not committing a type I error. Since our t statistic is smaller than the critical value at 0.01 (CV = 3.25) we can say that the probability of obtaining this t score, if the null hypothesis is true lies somewhere between 0.01 to 0.02.
2. The second example dealt with a horticultural group, whose seasonality of diet was being studied. The anthropologist uses the mean number of animal species consumed

during the wet season (10) to propose the μ of the null hypothesis. Thus, H0: $\mu = 10$ and H1: $\mu \neq 10$. As usual, a two-tailed test at a 0.05 level will be used.

The researcher investigates $n = 20$ households, and determines that the mean number of animal species consumed during the dry seasons was $\bar{Y} = 9$ with a standard deviation $s = 3.5$. The t score is computed as follows: $s_{\bar{Y}} = \frac{3.5}{\sqrt{20}} = 0.78$ and $t = \frac{10-9}{0.78} = 1.28$. At $\alpha = 0.05$ with df $= 20 - 1 = 19$, the critical value is 2.093. Thus, the investigator fails to reject the null hypothesis: the community does not appear to consume a significantly different number of animal species during the dry season from the number consumed during the wet season.

3. The last example concerns the number of words to refer to soil quality by males and females in a community. The parametric mean of the null hypothesis is proposed based on the number of words used by males. Thus, H0: $\mu = 4$. H1: $\mu \neq 4$. A sample of size $n = 30$ females is researched, and its mean number of words is determined to be $\bar{Y} = 7$, with a standard deviation $s = 1.8$. A two-tailed test with an α level of 0.05 will be used. The t score is computed as follows: $s_{\bar{Y}} = \frac{1.8}{\sqrt{30}} = 0.33$. $t = \frac{7-4}{0.33} = 9.1$. For df $= 30 - 1 = 29$ and at a level of 0.05 the critical value is 2.045. In fact, the t score is much greater than the critical value at 0.001 (3.659). Thus, the null hypothesis is rejected, and the alternative hypothesis that males and females in this community use different number of words for soil quality is accepted. What is the probability of committing a type I error in this situation? Less than 0.0001.

4.3.8 Reporting hypothesis tests

Different disciplines and different journals within the same discipline have different standards on how results of statistical tests should be reported. However, there is a minimum of information that should always be included when reporting statistical results: the value of the test statistic (say the t score), the degrees of freedom or the sample size, the probability associated with such a statistic (better known as the ***p* value**), and the conclusion reached about the hypothesis. How this information is written depends on the writer and the journal or book. For example, the researcher may report the results of the third example of the previous section as $t = 9.1$, df $= 29$, $p < 0.0001$, or as $t_{(29)} = 9.1, p < 0.0001$, or as $t_{(\mathrm{df}=29)} = 9.1^{***}$. The latter example illustrates the common use of asterisks, in which $^{*} = 0.05$ (but not significant to the 0.01 level), $^{**} = 0.01$ (but not significant to the 0.0001 level), and $^{***} = 0.0001$. Another frequently used convention is that if a t score (or an F ratio, or a X^2 statistic, as we will see later) is less than one in value, it is always not significant, no matter the degrees of freedom. In that situation, researchers simply report, say, $t = 0.8$, ns (for non-significant).

If the statistical analysis is done with computers, then the exact probability of a statistic such as a t score is known. That is, instead of knowing that the probability associated with a specific score is between 0.05 and 0.01, it is known that the probability is exactly, say, 0.025. In that case, researchers use an equal sign (instead of a "<" sign) when reporting the probability. The use of computers has simplified greatly what can be a laborious and error-prone task: looking up in a statistical table to find out if the results are significant. Since the computer prints the statistic, the degrees of freedom, and the p value associated with the statistic, there is no need to use the statistical table: if the probability is less than or equal to 0.05, then the statistic is significant.

Whatever specific notation researchers use, they must at some point of the paper specify: (1) what hypothesis is being tested, (2) what statistical test is being used, (3) the results of the statistical test, and (4) the conclusion reached about the hypothesis.

4.3.9 The classical significance testing approach. A conclusion

The approach to hypothesis testing presented in this textbook and in most other textbooks is the product of a specific historical development in scientific methodology. While Cohen refers to it as "... the accident of the historical precedence of Fisherian theory, its hybridization with the contradictory Neyman-Pearson theory ..." (Cohen 1992), Pollard refers to it as "... a somewhat inconsistent mixture of the two approaches ..." (Pollard 1986). Pollard (1986) goes further in indicating common mistakes in the application of the classical significance testing approach. He notes that if we use a 5% alpha level, and we accept the null hypothesis, this does not mean that the null hypothesis has a 95% chance of being true. Of course this is correct: a hypothesis is true or false, it does not have a probability of being true or false. In the same manner, when we reject a null hypothesis it is good practice to remind ourselves that we could be committing a type I error. Therefore, both α and β are probabilities of committing errors, they are not the probabilities of a null hypothesis being true or false.

A lack of consideration of the importance of both types of statistical errors is another issue which shows that the classical significance testing approach may be misapplied (Cohen 1992; Pollard 1986). If the data analyst is only concerned about rejecting the null hypothesis she loses track of how serious a type II error is. In reality, we need to consider the real-life implications of what we are doing: is it more serious to reject a true null hypothesis than to fail to reject a false null hypothesis? This brings us to another point: we need to remember that statistical significance does not mean real-life significance (Pollard 1986). All in all, the reader is reminded to try to do a power analysis when possible, setting up an alpha level which is relevant for the research project (which may be different from the usual 0.05 level) and interpreting a null hypothesis rejection or failure to reject appropriately.

4.4 Chapter 4 key concepts

- Discuss different approaches to estimation and hypothesis testing.
- Estimation.
- Confidence limits and confidence intervals.
- The principles of hypothesis testing.
- Alternative and null hypotheses.
- Significance level.
- Type I error.

- Type II error.
- Statistical power.
- One- and two-tailed hypothesis tests.
- Independence of variates.
- Non-parametric test.
- t distribution.
- In research papers in your area of studies, do you see the misapplications noted in section 4.3.9?

4.5 Chapter 4 exercises

1. Create a research problem of interest to you and describe your null and alternative hypotheses.
2. What is estimation? Explain the difference between point and interval estimation.
3. What is the difference between a t score and a z score?
4. Given that in a hypothesis test the H0: $\mu = 0$, the H1: $\mu = 2$, and the standard deviation is $\sigma = 1$, compute the following:
   a. If $n = 5$, what is the test's power?
   b. If $n = 20$, what is the test's power?
   c. If H0: $\mu = 0$, the H1: $\mu = 3$, the standard deviation is $\sigma = 1$, and $n = 5$, what is the test's power?
   d. If H0: $\mu = 0$, the H3: $\mu = 2$, the standard deviation is $\sigma = 1$, and $n = 20$, what is the test's power?
5. Test the hypotheses that these sample means belong to the population proposed by the H0. Use both a z score and a t score.
   a. For the z score: H0: $\mu = 18$, $\sigma = 5$, $n = 90$, $\bar{Y} = 20$. For the t score: H0: $\mu = 18$, $s = 5.3$, $n = 90$, $\bar{Y} = 20$.
   b. For the z score: H0: $\mu = 178$, $\sigma = 45$, $n = 300$, $\bar{Y} = 172$. For the t score: H0: $\mu = 178$, $s = 180$, $n = 300$, $\bar{Y} = 172$.
6. Find out how statistical tests are usually reported in a journal in your field of study.

5 The difference between two means

In the previous chapter we covered the topic of hypothesis testing with z and t scores. The hypothesis we tested in both cases was that a particular sample $\bar{Y}$ was obtained from a population with mean μ. We discussed that we could use z scores only if we knew the population's σ (a very rare situation), and that in most cases we use a t score which is computed using the sample's standard deviation. Notice that in either case we are comparing a sample mean with the parametric mean of a hypothetical population. This is not a very common experimental design. Instead, it is more frequent to compare two sample means with each other, testing the null hypothesis that both were collected from the same population. Such a hypothesis test is covered here, specifically applied in the following cases:

1. The difference between two independent sample means (e.g. the mean weight of two samples of children independently collected in two different locales of a community).
2. The comparison of an observation with a sample (e.g. an arrowhead with a sample of arrowheads).
3. The comparison of two paired groups (e.g. the weight of a group of individuals before and after a smoking-cessation program). The reader should note that in principle, all of these tests could be done using z scores if we had knowledge of the population parameters. Since this is rarely the case, we test these hypotheses using t scores.

5.1 The un-paired *t* test

This is a popular and frequently used statistical test, commonly referred to as "**the *t* test**." In reality, this test is a special case of an analysis of variance (ANOVA), which is covered in the next chapter. However, the t test is so popular that it deserves to be in its own chapter. This test is also frequently known as "Student's t test" because the t distribution was first proposed by Gosset in a paper which he authored under the pseudonym "Student" (Fienberg and Lazar 2001).

The t test is used if a researcher needs to compare two samples to determine if they differ significantly. Examples in anthropology are the mean age at marriage of two ethnic groups, the mean length of projectile points, the mean height of two groups of children, etc. *The null and alternative hypotheses* in a two-tailed test are H0: $\mu_1 = \mu_2$, and H1: $\mu_1 \neq \mu_2$. A one-tailed test should only be done (as always) when the researcher knows, or is interested in testing only that one group has a greater or smaller mean than the other. In a one-tailed test, the null and alternative hypothesis are H0: $\mu_1 \geq \mu_2$, and H1: $\mu_1 < \mu_2$, or H0: $\mu_1 \leq \mu_2$, and H1: $\mu_1 > \mu_2$, depending on the direction of the treatment effect.

The problem when comparing two sample means is that we need to agree on how different two means have to be for us to decide that they were not sampled from the same population. As we learned in chapter four, means of sample size n will be approximately normally distributed around the parametric mean. Thus, some means from the same population may be different enough from each other for us to decide (in error) that they were not sampled from the same population. On the other hand, some means do not differ at all and their difference is 0, which is the mean of the distribution of sample-mean differences. The more different the two samples are, the larger the difference between the means will be. As the difference increases in value it will be farther apart from the mean of sample differences ($\mu = 0$), and it will have a lower frequency because it will be located towards the tails of the distribution. What we are discussing here is nothing more than a normal distribution, whose properties we are familiar with. If a difference falls within the customary 95% acceptance region, it will be declared to be non-significantly different from the mean of zero, and the null hypothesis will be accepted. If the difference falls in the rejection region, usually distributed equally in both tails, then the difference will be declared to be significantly different from the mean of zero, and the null hypothesis will be rejected. This rejection will be made with the acknowledgment that such a decision could be in error, and will be accompanied by the probability (p value) that we have committed a type I error.

It is helpful to think of the t test in terms of dependent and independent variables. The dependent variable is that whose behavior we want to study, such as the weight of children, the size of the arrowheads, or the family income. The independent variable is that under the control of the researcher, and it is the one which is used to categorize the dependent variable into groups: the two community centers where we collected the weight of children, the two sites where we collected the arrowheads, and the two neighborhoods where we collected family income. If we think of the t test in terms of dependent and independent variables it will be easier to understand the ***t* test as a form of regression analysis**. When we propose to determine whether children from two community centers differ in their weight we are trying to explain the reasons for the variation of the dependent variable (weight) in terms of the independent variable (the two community centers and their socioeconomic realities). We will designate the dependent variable with a Y and the independent variable with an X. In this example the Y is a continuous numerical variable, while the X is a discrete variable with two levels: center A and center B. In regression analysis we will refer to the latter as a **dummy independent variable**. Looking at this experimental design with a regression analysis view allows us to predict that the weight of children will be: $\hat{Y}_{ij} = \mu + \alpha_i + \epsilon_{ij}$, where $\hat{Y}_{ij}$ = the predicted weight for the jth child who is in the ith group ($\hat{Y}$ is referred to as **"Y-hat"**), μ = the mean of all children in the study regardless of the group to which they belong, α_i is **the treatment effect** of being in group i^{th}, and ε_{ij} is the normal random variation found in the ith group which is measured in the jth child. In a regression-analysis nutshell, a t test will allow us to see if the treatment effect α_i is significant so that the children of the two community centers differ from each other and cannot be said to belong to the same population whose mean is μ. It is a regrettable convention that the treatment effect is denoted with the symbol α which is usually taken to mean the probability of committing a type I error.

The formula for the t score to test the null hypothesis that both samples were obtained from the same population is very similar to that of the t score we used in chapter four (to test that a sample was obtained from a population of mean μ). Both have a difference in the numerator, which in the t score we used previously was the difference between the sample mean $\bar{Y}$ and the population mean μ. In the t score we will use in this chapter, the difference is between that of the two $\bar{Y}$s and that of the two μs ($[\bar{Y}_1 - \bar{Y}_2] - [\mu_1 - \mu_2]$). Since according to the null hypothesis, the two samples were obtained from the same population, the difference between the two parametric means is 0. Thus, the second term ($\mu_1 - \mu_2$) is removed, and the numerator is only $\bar{Y}_1 - \bar{Y}_2$. The denominator of the t test to compare two samples is also very similar to that of the t score to compare a sample mean to a population mean. The reader recalls that in the latter case we obtained the denominator by dividing the sample's standard deviation by the square root of the sample size. We did this because the standard deviation of sample means depends on the sample size as well as on the variation of observations in the sample. In the t score we will use in this chapter we need a similar measure, but one that reflects not the standard error of means, but the standard error of *mean differences*. The **standard error of mean differences** $s_{\bar{Y}_1-\bar{Y}_2}$ incorporates the sample sizes (n_1 and n_2), and standard deviations (actually, the variances s_1^2 and s_2^2 are chosen for ease of computation). Lastly, to compare our t score with the t table, we need to use a different formula for calculating the degrees of freedom, one that takes into consideration the sample sizes of both groups so that $df = n_1 + n_2 - 2$. As usual, if the t score we calculate is greater than or equal to the critical value in the table, we reject the null hypothesis. Table 5.1 shows selected values of the t distribution.

Formula 5.1

The t score to test that two sample means were obtained from the same population.

$$t = \frac{\bar{Y}_1 - \bar{Y}_2}{\sqrt{\left[\frac{(n_1-1)s_1^2+(n_2-1)s_2^2}{n_1+n_2}\right]\left(\frac{n_1+n_2}{(n_1)(n_2)}\right)}}$$

with $df = n_1 + n_2$

Let us practice the computation of the t test with the data set called "projectile points in two sites," which has the length (in centimeters) of projectile points collected in two fictitious sites. Is there a significant difference in the mean length of the projectile points of the two sites? The data set is in Excel format at the book's website.

Table 5.1 Selected critical values of the *t* distribution.

	Proportion in one tail			
	0.05	0.025	0.01	0.005
	Proportion in two tails			
df	0.10	0.05	0.02	0.01
1	6.314	12.706	31.821	63.657
5	2.015	2.571	3.365	4.032
6	1.943	2.447	3.143	3.707
7	1.895	2.365	2.998	3.499
9	1,833	2.262	2.821	3.250
10	1.812	2.228	2.764	3.169
13	1.771	2.160	2.650	3.012
15	1.753	2.131	2.602	2.947
20	1.725	2.086	2.528	2.845
25	1.708	2.060	2.485	2.787
30	1.697	2.042	2.457	2.750
40	1.684	2.021	2.423	2.704
60	1.671	2.000	2.390	2.660
120	1.658	1.980	2.358	2.617
∞	1.645	1.960	2.326	2.576

We compute the following information:

$$n_1 = 30,\ \bar{Y}_1 = 7.9,\ s_1^2 = 3.67.$$
$$n_2 = 30,\ \bar{Y}_2 = 9.9,\ s_2^2 = 3.67.$$

We follow the usual steps when testing a hypothesis:

1. We state the null and alternative hypothesis: H0: $\mu_1 = \mu_2$, H1: $\mu_1 \neq \mu_2$.
2. The alpha level will be 0.05. Since our degrees of freedom $(30 + 30 - 2 = 58)$ are not in the table we use the next lower value or df $= 40$. At the 0.05 level for a two-tailed test the critical value is 2.021. Thus, if our *t* score is greater (in absolute value, since the *t* distribution is symmetrical) than or equal to 2.021, we will reject the H0.
3. The sample has been collected and given to us.
4. The *t* score is computed and a decision is reached.

$$t = \frac{7.9 - 9.9}{\sqrt{\left[\frac{(30-1)^*3.67+(30-1)^*3.67}{30+30}\right]\left(\frac{30+30}{(30)(30)}\right)}} = \frac{-2}{\sqrt{\left(\frac{106.43+106.43}{60}\right)\left(\frac{60}{900}\right)}} =$$

$$t = \frac{-2}{\sqrt{(3.55)\,(0.06666)}}\frac{-2}{0.48646} = -4.11.$$

Since the *t* score we computed is greater than the critical value at $\alpha = 0.05\,(2.021)$ we reject H0. In fact, our *t* score is greater than the critical value at $\alpha = 0.02$ (2.423) and $\alpha = 0.01$ (2.704). Therefore, we reject at a probability even lower than 0.01 (which, as we should always keep in mind, is the probability that we are committing a type I error).

Practice problem 5.1

The data set called "cranial morphology" has different cranial measurements from a skeletal population (please see the Word document describing the data set for an explanation of the variable names). Each specimen has been assigned an F for female and an M for male. The assignation of sex was based on non-cranial data. Do male and female skeletons differ for the variable BNL?

We compute the following information for the variable BNL:

$$n_F = 53,\ \bar{Y}_F = 98.42,\ s_F^2 = 13.860.$$
$$n_M = 66,\ \bar{Y}_M = 103.65,\ s_M^2 = 21.058.$$

We follow the usual steps when testing a hypothesis:

1. We state the null and alternative hypothesis: H0: $\mu_1 = \mu_2$, H1: $\mu_1 \neq \mu_2$.
2. The alpha level will be 0.05. Since our degrees of freedom ($53 + 66 - 2 = 117$) are not in the table we use the next lower value or df = 60. At the 0.05 level for a two-tail test the critical value is 2.0. Thus, if our t score is greater (in absolute value, since the t distribution is symmetrical) than or equal to 2.0, we will reject the H0.
3. The sample has been collected and given to us.

$$t = \frac{98.42 - 103.65}{\sqrt{\left[\frac{(53-1)^{*}13.86+(66-1)^{*}21.058}{53+66}\right]\left(\frac{53+66}{(53)(66)}\right)}} = \frac{-5.23}{\sqrt{\left(\frac{720.72+1368.77}{119}\right)\left(\frac{119}{3498}\right)}} =$$

$$t = \frac{-5.23}{\sqrt{(17.5587)(0.0340)}}\frac{-5.23}{0.77266} = -6.77.$$

Since the t score we computed is greater than the critical value at $\alpha = 0.05$ we reject H0. In fact, our t score is greater than the critical value at $\alpha = 0.02$ (2.39) and $\alpha = 0.01$ (2.660). Therefore, we reject at a probability even lower than 0.01.

5.1.1 Assumptions of the un-paired *t* test

It is interesting that the assumptions of statistical tests tend to receive little attention in many texts, and that they are usually discussed after the test itself is discussed, even though the assumptions should be tested before the test itself is performed. In real life, researchers should not proceed to apply a t test without testing some fundamental assumptions about the data, because if the data violate the assumptions, the results of the test may be meaningless.

In the last chapter we discussed assumptions of several statistical tests which will only be briefly mentioned here. The assumptions which must be tested to perform the t test will receive more attention.

5.1.1.1 Random sampling

The *t* test assumes that the samples were collected giving each member of the population an equal chance of being selected.

5.1.1.2 Independence of variates

The assumption means that if an observation is collected, it will not influence which observation is collected next in the same sample. The assumption also extends to the other sample: both are independent from each other.

5.1.1.3 Normality of data

The assumption of normality is frequently not given much importance because most samples of size $n = 30$ are normally distributed. However, both PASW and SAS offer extremely easy ways of testing if data are normally distributed (noted below, under "computer resources"). Both of these computer packages allow the user to obtain tests which test the null hypothesis that the data are normal. If the probability associated with these tests is less than or equal to 0.05 the user rejects the null hypothesis of normality. The tests produced by both packages are the **Kolmogorov–Smirnov** (K-S) and the **Shapiro–Wilk** (S-W) tests (the former does not refer to a vodka brand nor the latter to a law office; instead both tests are named after their proponents). In addition, SAS produces two other tests (Cramer–von Mises and Anderson–Darling). Please note that although these tests are convenient, their statistical power has been doubted (Zar 1999) to such an extent that Zar proposes other normality tests not supported by either computer package. Indeed, it has happened to me that I get contradictory conclusions from these tests, that is, one accepts and the other one rejects the null hypothesis of normality (this is why I do agree with Zar that the results of these tests should be taken with caution). Let us test the normality of the variable BNL for the male and female skeletons from practice problem 5.1. The following results were obtained with SAS:

Tests for normality for females. BNL

Test	Statistic	*p* value
Shapiro–Wilk	W 0.962833	0.0979
Kolmogorov–Smirnov	D 0.134861	0.0173
Cramer–von Mises	W-Sq 0.114155	0.0745
Anderson–Darling	A-Sq 0.673948	0.0781

Tests for normality for males. BNL

Test	Statistic	*p* value
Shapiro–Wilk	W 0.980338	0.3782
Kolmogorov–Smirnov	D 0.101898	0.0879
Cramer–von Mises	W-Sq 0.076207	0.2336
Anderson–Darling	A-Sq 0.476533	0.2366

You can see that there is disagreement among the tests as to whether BNL is normally distributed in females, since one of the tests has a probability of 0.01. However, since the other tests do suggest some minor departure from normality as their probability is 0.09 and 0.07 I would take these results as a strong suggestion that the female subsample is not normally distributed for the BNL variable.

In a case such as this, to help you understand what is happening with the distribution of your data, perhaps it is a good idea to determine if the data suffer from **skewness** or **kurtosis**. You recall that a data set is skewed to the right if the data observations are clustered in the left side of the distribution, while a data set is skewed to the left if the data are clustered on the right side of the axis (it helps me to think that the long tail is pointing to the area to which the data are skewed). Skewness is measured by a statistic whose derivation I will not cover here, called g_1. If g_1 is positive then the distribution is skewed to the right, while if g_1 is negative then the distribution is skewed to the left. Kurtosis in a data set is measured by another statistic called g_2, which if negative indicates platykurtosis, and if positive indicates leptokurtosis. In a normal distribution, the parametric value of g_1 and g_2 (γ_1 and γ_2) is 0. It seems that it would be simple to test the null hypothesis that both statistics computed from the data set are equal to 0, but this is not the case because any tests of such hypothesis must be done on large data sets. Indeed, it is reassuring that neither PASW nor SAS provide such a statistical test but rather provide a standard error of the statistic. If the user is sure that he has a large enough data set, then he can compute a t score in the usual manner where $t = \frac{g_1 - \gamma_1}{s_{g_1}}$ or $t = \frac{g_2 - \gamma_2}{s_{g_2}}$, using df $= \infty$. Therefore a t score equal to or greater than 1.96 (the critical value at the 0.05 probability level for a two-tailed test) should lead to rejection of the null hypothesis that H0: $g_1 = 0$ or $g_2 = 0$. According to several PASW and SAS manuals, the user should not compute this t score unless his sample size is at least 150.

Practice problem 5.2

Let us compute the t scores for both g_1 and g_2 by sex of the skeletons and for BNL. Keep in mind that γ_1 and γ_2 have values of 0.

Females. BNL				
Skewness	−0.559	Std. Error	0.327	$t = \frac{g_1 - \gamma_1}{s_{g_1}} = \frac{-0.559}{0.327} = -1.71$
Kurtosis	−0.202	Std. Error	0.644	$t = \frac{g_2 - \gamma_2}{s_{g_2}} = \frac{-0.202}{0.644} = -0.31$
Males. BNL.				
Skewness	−0.219	Std. Error	0.306	$t = \frac{g_1 - \gamma_1}{s_{g_1}} = \frac{-0.219}{0.306} = -0.716$
Kurtosis	0.142	Std. Error	0.604	$t = \frac{g_2 - \gamma_2}{s_{g_2}} = \frac{0.142}{0.604} = 0.235$

All t scores are less than the critical value of 1.96 and fall in the acceptance region. Therefore, our data set does not suffer from significant skewness or kurtosis. Please note that these tests were done on a sample size below the recommended size of $n = 150$.

Table 5.2 Selected critical values of the maximum *F* ratio ($\alpha = 0.05$).

	a (number of groups)			
$n-1$	2	3	4	5
10	3.72	4.85	5.67	6.34
15	2.86	3.54	4.01	4.37
20	2.46	2.95	3.29	3.54
30	2.07	2.40	2.61	2.78
60	1.67	1.85	1.96	2.04
∞	1.00	1.00	1.00	1.00

5.1.1.4 Homoscedasticity

The reader recalls that a *t* score that tests if two samples are significantly different has as its denominator the **standard error of mean differences** or $s_{\bar{Y}_1-\bar{Y}_2}$. This statistic incorporates the variances and sample sizes of both groups, and in effect combines them. Since, according to the null hypothesis, both samples came from the same population, they should have homogeneous variances (a condition known as **homoscedasticity**). If the variances are not homogeneous (**heteroscedasticity**) then it is possible for a null hypothesis to be rejected not because the sample means truly differ from each other, but because the sample variances do.

Fortunately both PASW and SAS will test this assumption by default and compute an **approximate *t* test for unequal variances** which the user can use should the test for homogeneity of variances be significant. Although the formula for these *t* tests is not complicated, the formula for computing their degrees of freedom is, and it may actually result in degrees of freedom which are not integers. Therefore, do not be surprised if the computer package prints a df such as 37.6.

In sum, the analyst should: (1) test the null hypothesis that H0: $\sigma_1^2 = \sigma_2^2$ and (2) use whichever *t* score is appropriate. Thus, if a researcher has access to either of these packages, the testing of the homoscedasticity assumption is part of the *t* test itself. If investigators do not have access to a computer package that does this, they must test the assumption by hand. A commonly used test is the F_{max} test, in which the larger variance is divided by the smaller variance, and the ratio (known as the F_{max} ratio) compared with a table of critical values (Table 5.2). Two degrees of freedom are needed to use this table: the first is the number (a) of groups compared ($a = 2$ in a *t* test), and the second is $n - 1$, where n is the size of the smaller sample. If the F_{max} statistic is greater than or equal to the critical value then the null hypothesis of homoscedasticity is rejected.

Practice problem 5.3

Test the assumption of homoscedasticity using the example from practice problem 5.1.

$$n_F = 53,\ \bar{Y}_F = 98.42,\ s_F^2 = 13.860.$$

$$n_M = 66,\ \bar{Y}_M = 103.65,\ s_M^2 = 21.058.$$

Our degrees of freedom are $a = 2$ and $n - 1 = 53 - 1 = 52$. The closest degrees of freedom are 30. The critical value is 2.07. The null hypothesis is H0: $\sigma_1^2 = \sigma_2^2$. We divide the larger variance by the smaller to obtain the F_{max} statistic: $F_{max} = \frac{21.058}{13.860} = 1.519$. We do not reject the null hypothesis. Our data do not violate the assumption of homogeneity of variances.

5.2 The comparison of a single observation with the mean of a sample

Students may sometimes be in a situation in which they have a single variate (a projectile point, a bone, etc., symbolized as y_i) of unknown origin. If the researchers have one or more samples from which the observation could have come, then they can establish with a t score if the observation is or is not significantly different from the sample mean. If the observation is not significantly different from the sample mean, the researchers have not, of course, proven that the observation belongs to the population from which the sample was obtained. But at least they know that the observation is not significantly different from the sample mean. The null hypothesis is that the μ of the population from which the observation was sampled is the same as the μ of the population from which the sample mean was obtained. Thus, H0: μ = the value estimated by the sample mean. The t score used to compare one observation with a sample mean is very similar to that presented in the previous section, and it is shown in formula 5.2.

Formula 5.2

The *t* score to compare a single observation with a sample mean.

$$t = \frac{y_i - \bar{Y}}{(s)\left(\sqrt{\frac{n+1}{n}}\right)}, \quad \text{df} = n = 1$$

Let us apply this formula by supposing that we found a projectile point of length y = 5.5 close to two samples, whose means, standard deviations, and sample sizes are:

SITE	n	Mean	Std. Dev.
1	8	4.40	0.90
2	7	5.90	0.63

Let us compute a t score to compare our observation with both samples:

1. H0: μ = 4.4.

$$t = \frac{5.5 - 4.4}{(0.9)\left[\sqrt{\frac{8+1}{8}}\right]} = \frac{1.1}{(0.9)[1.06]} = 1.15, \; p > 0.05.$$

2. H0: $\mu = 5.9$.

$$t = \frac{5.5 - 5.9}{(0.63)\left[\sqrt{\frac{7+1}{7}}\right]} = \frac{-0.4}{(0.63)[1.07]} = -0.59, ns.$$

In this example, the observation is not significantly different from either sample. At df = 7 the critical value is 2.365, and at df = 6, the critical value is 2.447.

5.3 The paired *t* test

One of the assumptions for the un-paired *t* test was that sampling be random. That is, both samples were obtained randomly from their populations. In a paired comparison, however, we frequently work with one sample instead of two, comparing the mean of the sample at two different times. Or if we work with two distinct samples, they are paired or matched for variables said to be controlled. Examples of this experimental design are the comparison of the mean weight in a group of individuals before and after a diet program, or the mean height in a group of children taken a year apart, or the achieved fertility in two groups of women matched for education, social class, and contraceptive use but not matched for religion.

In a paired comparison, we cannot use the *t* scores we used in the previous section. Even if we are working with two distinct samples (as in the example of the two groups of females), what interests us most is the magnitude of the differences between the two groups. Indeed, in a paired comparison we do not work with the data themselves, but instead create a new variable, the ***differences*** (*D*) between the observations of both groups. *The null hypothesis* states that the parametric mean of the population of differences from which our sample of differences was obtained has a value of 0. Thus: H0: $\mu_D = 0$. The alternative hypothesis is usually two-tailed. The structure of this *t* score is very similar to those previously discussed: in the numerator we are going to have a difference, namely, the difference between the mean sample difference ($\bar{D}$) and the difference between the parametric means of the populations from which our samples were obtained ($\mu_1 - \mu_2$). Since, according to the null hypothesis, the two populations are the same, $\mu_1 - \mu_2$ is 0 and the numerator consists only of $\bar{D}$. The denominator consists of a very familiar term, namely, the **standard error of the differences** or $s_{\bar{D}}$, $= \frac{s}{\sqrt{n}}$, where *s* is the standard deviation *of the differences* and *n* is the sample size. The formula for the paired *t* test is shown in formula 5.3.

Formula 5.3

The *t* score for paired comparisons.

$$t = \frac{\bar{D}}{s_{\bar{D}}}, \text{ where}$$

$$s_{\bar{D}}, = \frac{s}{\sqrt{n}}, \quad \text{df} = n - 1$$

Let us say that an anthropologist has been hired to determine if the test scores of a group of $n = 10$ children have changed significantly after the introduction of a new teaching methodology. The data are listed below, as well as a column of differences between both measures. It does not matter how the differences are obtained. However, if the computations are to be done by hand, it is easier to handle them if we obtain the least number of negative differences. The usual steps in hypothesis testing are followed:

1. The null and alternative hypothesis are stated: H0: $\mu_D = 0$ and H1: $\mu_D \neq 0$.
2. An alpha level of 0.05 is chosen. At df = 9, the critical value is 2.262.
3. The data are collected:

	Before	After	Difference (before – after)
1.	80	95	−15
2.	70	73	−3
3.	60	55	5
4.	70	75	−5
5.	90	87	3
6.	65	76	−11
7.	70	65	5
8.	55	50	5
9.	80	82	−2
10.	95	95	0

4. The t score is computed and a decision is reached.

$$\bar{D} = -1.8, s = 6.96, s_{\bar{D}} = \frac{6.96}{\sqrt{10}} = 2.2, t = \frac{-1.8}{2.2} = 0.818$$

Since any t score whose value is less than 1 is known not to be significant, we accept the null hypothesis, and conclude that the new teaching methodology has not changed the test results.

Practice problem 5.4

An anthropologist interviews 14 males about their reproductive history. One question concerns how old the male was when he had his first child. In an attempt to test the reliability of the subjects' answers, the researcher re-interviews the subjects three months later. Test the hypothesis that there is no difference in the answers.

1. The null and alternative hypotheses are stated. H0: $\mu_D = 0$ and H1: $\mu_D \neq 0$.
2. An alpha level of 0.05 is chosen. The critical value for df = 14 − 1 = 13 is 2.16.

3. The data are collected, and are listed below:

	First interview	Second interview	Difference (first − second)
1.	18	18	0
2.	20	20	0
3.	25	15	10
4.	20	19	1
5.	30	29	1
6.	20	15	5
7.	23	23	0
8.	27	27	0
9.	23	16	7
10.	30	30	0
11.	25	24	1
12.	31	31	0
13.	26	20	6
14.	22	22	0

4. The *t* score is computed, and a decision is reached:

$$\bar{D} = 2.21,\ s = 3.33,\ s_{\bar{D}} = \frac{3.33}{\sqrt{14}} = 0.89,\ t = \frac{2.21}{0.89} = 2.48,$$
$$\text{df} = 14 - 1 = 13,\ p < 0.05.$$

The null hypothesis of no differences is rejected: the subjects provided significantly different answers to the interview.

5.3.1 Assumptions of the paired *t* test

The paired *t* test has the following assumptions:

1. *Random sampling.* The paired *t* test assumes that the samples were collected giving each member of the population an equal chance of being selected.
2. *Independence of variates.* The assumption means that if an observation is collected, it will not influence which observation is collected next in the same sample. Of course, the two samples being compared are not independent from each other.
3. *The data are normal.* The assumption of normality affects the differences between the two measures, not the measures themselves. Therefore, the differences must be tested to determine if they have a normal distribution.

The un-paired and paired *t* tests in anthropology. According to Pakhretia and Pirta (2010) it is important to study general behavioral patterns of herding animals kept by transhumant human groups. Specifically, Pahretia and Pirta (2010) are interested in determining how goats and sheep differ in their social, feeding, and maternal behaviors in herds held by the Gaddis in the northwest Himalayas. These authors recorded the number of grazing and social behaviors observed in an animal during a 10-minute session. In their tables 2 and 3 the authors (Pakhretia and Pirta 2010) present the mean number of behaviors observed in the two species, and the *t* ratios and *p* values used to compare the means. The authors conclude that while goats graze more frequently at shoulder level and on their hind legs, sheep graze head down more frequently. In addition sheep mothers approached their offspring significantly more frequently than did goat mothers.

Forensic anthropologists frequently use the length of long bones to estimate the stature of a person whose skeletal remains are being researched. But, can the right and left long bones be used with equal confidence to estimate stature? Or is bilateral asymmetry in long bone length a necessary consideration? Krishan and colleagues (2010) apply the paired *t* test to determine if limb dimensions differ between the right and left limbs in the same person. They worked with adult males from a North Indian community who are engaged in strenuous physical activity on a daily basis. The authors conclude that if stature is estimated with bones from one side using a formula developed for bones from the other side then there is a high possibility of obtaining erroneous results. Forensic anthropologists should determine whether the bone with which they are working is right or left, and use side-appropriate formulae for estimating height (Krishan *et al.* 2010).

5.4 Chapter 5 key concepts

- The use of *z* scores vs. *t* scores. Why is it unlikely that you will use *z* scores in your research?
- Un-paired *t* test: when is it appropriate to use?
- Assumptions of the un-paired *t* test.
- Define homoscedasticity and heteroscedasticity.
- Define kurtosis and skewness.
- Explain departures from normality.
- The comparison of a single individual to a sample mean.
- Paired *t* test: when is it appropriate to use?
- Assumptions of the paired *t* test.

5.5 Computer resources

1. To obtain normality tests in SAS you need to use Proc Univariate normal; this procedure will give you several normality tests plus measures of kurtosis and skewness.

2. For obtaining information about the skewness and kurtosis of a distribution I favor PASW over SAS because the former gives you both g_1 and g_2 and their standard errors, which allow you to compute the t score explained in the previous paragraph.
3. To test the normality of data in PASW you need to go through two different paths. To obtain measures of kurtosis and skewness you follow Analyze → descriptive statistics → descriptives → options OR Analyze → descriptive → frequencies → statistics. To obtain normality tests in PASW you follow Analyze → descriptive statistics → explore → plots → normality plots with tests.

5.6 Chapter 5 exercises

All data sets are available for download at www.cambridge.org/9780521147088.

1. Use the data set called "cranial morphology." Compare males and females with a t test for all the variables not compared in this chapter, testing for normality and homoscedasticity. In your conclusions, indicate which if any variables violated the assumptions, so that you can test again for equality in the chapter on non-parametric tests.
2. Use the data set called "Education in children of two genders and three religions." Use a t test to determine if the two genders differ in the mean number of years of formal education. Test the assumptions.
3. The data set called "Infant birth weight" shows data on maternal health and baby weight collected on women who smoked and who did not smoke during pregnancy. Compare these variables between women who smoked and women who did not smoke.
4. The data set called "Number of idols in a community" shows the number of traditional idols displayed in households visited a year apart (see Word file explaining the research project). Has there been a significant change in the number of idols displayed? Test the assumption of normality of the differences.

6 The analysis of variance (ANOVA)

In this chapter we deal with a fundamentally important technique to analyze experimental data. The analysis of variance (ANOVA from now on) is due to Fisher (in whose honor the *F* distribution – used to test the ANOVA's H0) is so named (Fisher 1993; Zabell 2001). In experimental, lab-based fields, in which researchers have much control over experimental and control groups, the applications of ANOVA are almost endless. It is for this reason that textbooks written for biology and psychology students include exhaustive chapters on many ANOVA designs (Pagano and Gauvreau 1993; Rosner 2006; Sokal and Rohlf 1995; Zar 1999). In anthropology we are more limited by the fact that we are unable to control with much precision different levels of the independent variable and have (in general) less control over our experiment. My treatment of ANOVA is thus rather limited when compared with textbooks in biology, psychology, or public health.

6.1 Model I and model II ANOVA

It is easy for students to be overwhelmed when they read that there are model I, model II, and mixed-model, and one-way and two-way ANOVAs. Let us clarify what models I and II are. Mixed-model ANOVAS will be discussed when we cover two-way ANOVAs, in section 6.5. The experimental design of a **model I ANOVA** is similar to that of an un-paired *t* test. Indeed, I mentioned that the *t* test should be understood as a special form of the analysis of variance. In a one-way ANOVA we have more than two groups which have been subjected to different levels of the treatment. As we did for the *t* test, let us think of analysis of variance in terms of dependent (Y) and independent (X) variables. The dependent variable is that whose behavior we want to study such as the number of calories consumed by children. The dependent variable (Y) is usually a continuous numeric variable or a discontinuous numeric variable with a broad range of possible outcomes. In other words, you can do an ANOVA with a dependent discontinuous numeric variable such as the number of cows owned by families because the number can vary from 0 to many; if the Y takes only two values (such as "yes" and "no" or 0 or 1) or a few values (1–3) then you would need to analyze the data with a different statistical technique. The independent variable is that under the control of the researcher, and it is the one which is used to categorize the dependent variable into groups such as the children's socioeconomic status, a family's clan, etc. The independent variable can be seen as the grouping variable, and it is called "factor" by some computer packages.

When we think of model I ANOVA in terms of dependent and independent variables it is easier to understand that ANOVA is **a form of regression analysis**, in which we seek to explain the dependent variable as follows: $\widehat{Y}_{ij} = \mu + \alpha_i + \varepsilon_{ij}$, where $\widehat{y}_{ij}$ = the predicted y for the jth observation which is in the ith group, μ = the mean of all observations in the study regardless of the group to which they belong, α_i is the treatment effect of being in group i^{th}, and ε_{ij} is the normal random variation found within the ith group which is measured in the jth observation. In a regression-analysis nutshell, an ANOVA will allow us to see if the treatment effect α_i is significant so that the observations differ by group and cannot be said to belong to the same population whose mean is μ. As I noted in the previous chapter the **treatment effect** is denoted with the symbol α. In sum, *in a model I ANOVA the researcher has an obvious and repeatable treatment effect.* This experimental design allows the researcher to do another experiment (say record the calories consumed by children in another city and divide the children by socioeconomic status again) to replicate the first experiment. This is perhaps the most common type of ANOVA seen in anthropology. Possible questions are: do the family incomes differ in villages which are devoted to pastoralism, commerce, or to agriculture? Does fertility differ in women of three different religions? Do males of different ethnic groups differ in the years of education they receive? Does femur width differ in people of different occupations?

In a **model II ANOVA** there is no treatment effect. Instead, the researcher works with samples that have not been subjected to a treatment but have been obtained from **random sources** *out of the researcher's control.* If those sources have affected the subjects, then subjects within each source would be more alike than those of different sources. The purpose of a model II ANOVA is to determine if there is a component of variance due to the source of the subjects.

For the most part the computations of both models do not differ. The only difference is that if the null hypothesis is rejected, in a model II the researcher proceeds to quantify how much of the total variation is due to the source. This is referred to as computing the added component of variation. Generally in anthropology, there is an interest in finding out if a treatment affects the subjects. Therefore, a model I is usually applied, and is the only model discussed in detail in this book.

6.2 Model I, one-way ANOVA. Introduction and nomenclature

The best way to explain a one-way ANOVA design is to contrast it with a two-way ANOVA design. The experimental design of a one-way ANOVA is similar to that of an un-paired t test except that we work with more than two groups (or factors). In a one-way ANOVA there is only one independent variable. Therefore the observations are arranged in groups (or factors) defined by one variable only. For example, we may divide the children according to neighborhood (Y = weight, X = neighborhood, where there are four neighborhoods) or according to SES (Y = weight, X = SES, where SES takes 3 values), and so forth. In clinical terms, there is only one treatment effect (neighborhood OR SES). In contrast, in a two-way ANOVA the data are divided according to two treatments: children could be divided according to gender and neighborhood, and individuals according to ethnicity and socioeconomic status. The difference between a one-way and a two-way **model I ANOVA** is that the first one has one, and the second one has two treatment effects.

The *null hypothesis* of a model I, one-way ANOVA states that all samples were obtained from the same population. Therefore, the mean of their parametric population is one and the same. Thus H0: $\mu_1 = \mu_2 = \cdots = \mu_a$ where $a =$ the number of samples being compared. The alternative hypothesis states that the sample means differ and proposes that a treatment has affected at least one of the samples in such a way that the mean of that sample is unlikely to have been obtained from the same population from which the other samples were obtained.

Although the ANOVA nomenclature varies among textbooks, it is fairly consistent and easily followed. In this book, we will test *the null hypothesis* H0: $\mu_1 = \mu_2 = \cdots = \mu_a$ with ***a* samples** (*k* is frequently used instead of *a*). Each sample size will be referred to as n_i, with i having values from 1 to a. The total number of subjects is referred to as $\sum n_i$ (some texts use N instead, although this is misleading since N is the population size). As we learned in the previous chapter, the degrees of freedom (df) are computed as $n_i - 1$. To refer to the sum of variates in each sample, we will use $\sum Y_i$ (where the subscript refers to the group), and to refer to the sum of variates across all samples, we will use $\sum\sum Y$. In the same manner, if we need the sum of the squared variates per sample, we will refer to it as $\sum Y_i^2$ and if we need the sum of the squared variates for all observations across samples we will refer to it as $\sum\sum Y^2$. For each sample, we can compute the sum of squares ($\text{SS} = \sum Y_i^2 - \frac{(Y)_i^2}{n_i}$). As always, if the sum of squares is divided by the degrees of freedom, a variance is obtained ($S^2 = \frac{\sum Y_i^2 \frac{(Y)_i^2}{n_i}}{n_i - 1}$). In ANOVA nomenclature, however, variances are referred to as mean squares (MS), for reasons explained below.

I like to explain the procedure of a model I ANOVA as the procedure of slicing a pie. The first step in analysis of variance is to estimate the total amount of variation using all subjects without considering group membership. Therefore, we estimate a total sum of squares (where **SS total** $= \text{SS}_\text{Y}$). The total SS is similar to the entire pie which we wish to slice. We then proceed to partition this sum of squares into two components: the **SS error** or **SS within** and the **SS among**. By definition the SS total = SS within + SS among. Thus, the sums of squares are additive (in the same way in which all slices of the pie together equal the entire pie). The SS among quantifies the dispersion of sample means around the mean computed using all observations (the grand mean). The SS within quantifies the amount of variation found within samples, which can result from the normal variation found in any sample. The reason the SS within is also referred to as SS error is as follows: the best estimate we can make of the expected value of any observation within a sample is the sample mean. The variation around this sample mean can be seen as error, not in the sense of a mistake, but in the sense that data vary around the sample mean. The SS error quantifies this variation.

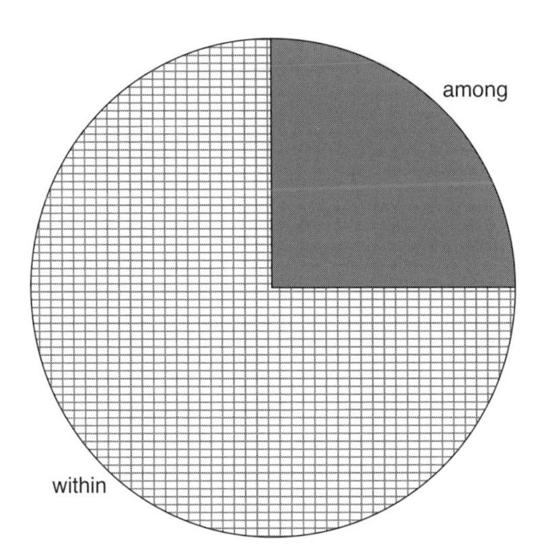

Figure 6.1 A small F ratio is obtained if there is more variation within than among groups.

After the three sums of squares are computed (SS total, SS within, and SS among), we can compute variances by dividing the sums of squares by degrees of freedom. The reason variances are not called variances in ANOVA is that the samples are not randomly obtained from the population proposed by H0. On the contrary, the groups have been subjected to a planned treatment. Thus, we refer to the quotient of the SS divided by the degrees of freedom as **mean squares** (virtually the mean squared deviation, because we divide by df instead of by n).

For the purposes of ANOVA, we only need to compute two mean squares: the within or error (MS within or MS error), and the among MS (MS among or MS factor, according to some computer packages). If the null hypothesis is correct, there is no treatment effect, and all samples were obtained from the same population. If that is the case, then subjects within groups are not more alike each other than they are to subjects from the other samples. That is, if the null hypothesis is true, there is as much variation within groups as there is among groups. If, on the other hand, the null hypothesis is to be rejected, then the subjects of at least one group are more alike each other than they are to subjects from other groups. If subjects from at least one group are alike and statistically different from subjects of other groups, then the mean of that group should be statistically significantly different from the means of the other samples.

We test the H0 by dividing the MS among by the MS within. Such division yields the ***F* ratio**, which is the statistic we use to decide if we reject or accept the H0. The following situations can occur: (1) If there is more variation within than among samples then we are dividing a small MS among by a larger MS within (we are dividing a small pie slice by a large one, Figure 6.1). The quotient in this case will be an F ratio less than 1, which is *always* non-significant. We would then accept the null hypothesis.

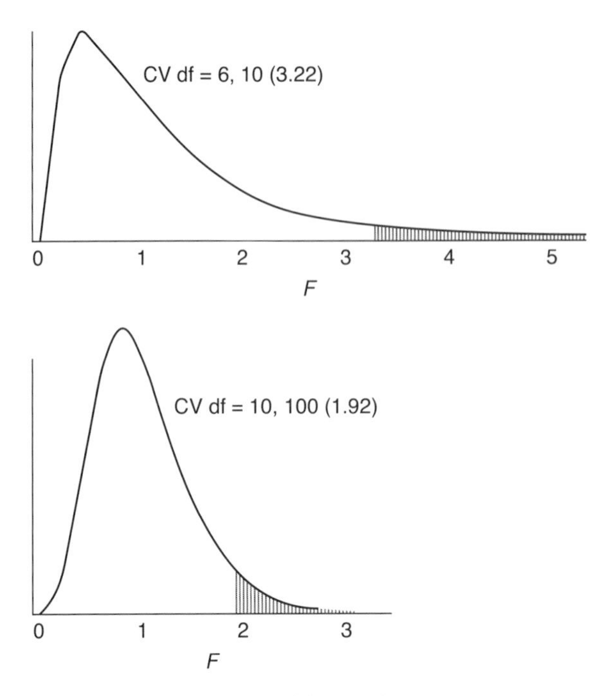

Figure 6.2 The distribution of the *F* ratio.

Notice that the smallest possible value the *F* ratio can take is that of 0: mean squares, which quantify variation, cannot possibly be negative, and the quotient of two positive numbers cannot be negative. This explains why the distribution of the *F* ratio is cut off at the value of 0 and is not symmetrical (see Figure 6.2).

(2) If the MS among and MS within are very similar in value then we divide a number by a very similar number (we divide the pie into two pieces of the same size, Figure 6.3). In that case, the quotient will be an *F* ratio very close to 1, and we will not reject H0.

(3) If the MS among is greater in value than the MS within (that is, there is less variation within groups than among groups because the treatment effect affected at least one sample) then we are dividing a large number by a small one (we divide a large pie slice by a small one, see Figure 6.4). The resulting *F* ratio will be large, and will fall in the distribution's rejection area, that is, it will be outside 95% of the distribution.

As in the *t* distribution, the distribution of the *F* ratio depends on the degrees of freedom. However, we need two degrees of freedoms to check the table of critical values (Table 6.1), since we are working with two variances. Table 6.1 lists on the top the degrees of freedom associated with the MS among (the numerator), and on the left side the degrees of freedom associated with the MS within (the denominator). If our *F* ratio is greater than or equal to the critical value, we reject the null hypothesis. It is customary to display all the information used to test the H0 in an ANOVA table (Table 6.2).

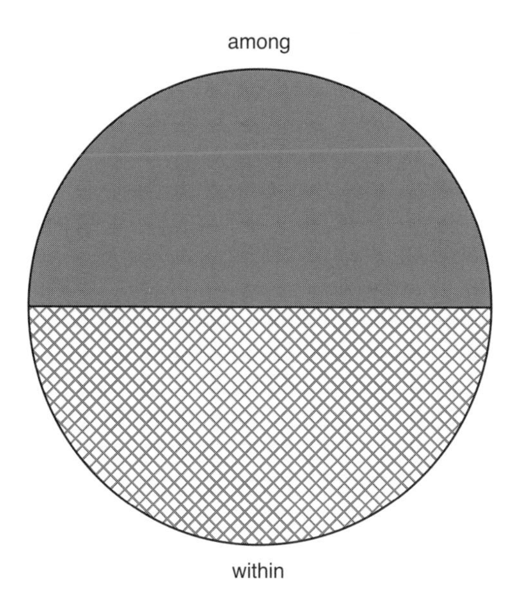

Figure 6.3 An F ratio close to 1 is obtained when the amount of variation within and among groups is similar.

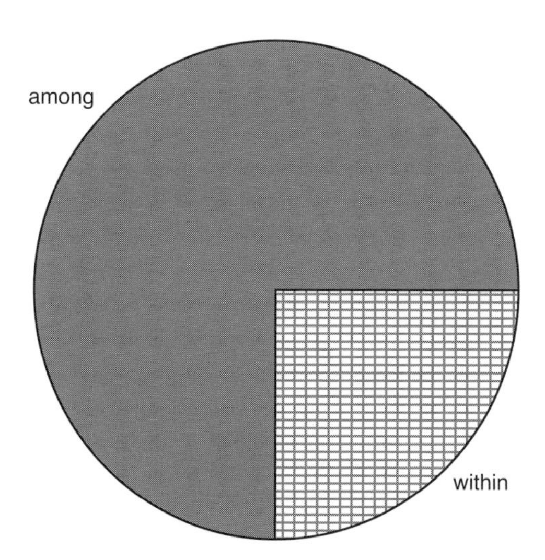

Figure 6.4 A large F ratio is obtained if there is more variation among than within groups.

The formulae we will use in this book can be used if the samples compared are of the same or different sizes and are more cumbersome than the formulae used for equal sample sizes. However, as presented in the table below, they can be readily computed from the data. All formulae are presented together instead of separately because they form a unit: the ANOVA table.

Table 6.1 Selected critical values of the *F* distribution.

Denominator degrees of freedom	Numerator degrees of freedom 1	2	3	4	5	6	7	...	10
10	4.96	4.10	3.71	3.48	3.33	3.22	3.14		2.97
	10.04	7.56	6.55	5.99	5.64	5.39	5.21		4.85
20	4.35	3.49	3.10	2.87	2.71	2.60	2.52		2.35
	8.10	5.85	4.94	4.43	4.10	3.87	3.71		3.37
30	4.17	3.32	2.92	4.69	2.53	2.42	2.34		2.16
	7.56	5.39	4.51	4.02	3.70	3.47	3.30		2.98
40	4.08	3.23	2.84	2.61	2.45	2.34	2.25		2.07
	7.31	5.18	4.31	3.83	3.51	3.29	3.12		2.80
60	4.00	3.15	2.76	2.52	2.37	2.25	2.17		1.99
	7.08	4.98	4.13	3.65	3.34	3.12	2.95		2.63
70	3.98	3.13	2.74	2.50	2.35	2.23	2.14		1.97
	7.01	4.92	4.08	3.60	3.29	3.07	2.91		2.59
100	3.94	3.09	2.70	2.46	2.30	2.19	2.10		1.92
	6.90	4.82	3.98	3.51	3.20	2.99	2.82		2.51

(The first entry is the critical value at the 0.05 and the second one is the critical value at the 0.01 level of significance.)

Table 6.2 Information presented in an ANOVA table.

Source of variation	df	Sums of squares (SS)	Mean squares (MS)	*F*	*p* value
Among groups	$a-1$	SS among	$\frac{\text{SS among}}{a-1}$	$\frac{\text{MS among}}{\text{MS within}}$	
Within groups	$\sum n - a$	SS within	$\frac{\text{SS within}}{\sum n - a}$		
Total variation	$\sum n - 1$	SS among + SS within			

Formulae 6.1

Formulae for the computation of ANOVA (one-way, model I).

Source of variation	df	Sums of squares (SS)	Mean squares (MS)	*F*	p
Among groups	$a-1$	$\sum \frac{(\sum Y_i)^2}{n_i} - \frac{(\sum\sum Y)^2}{\sum n}$	$\frac{\text{SS among}}{a-1}$	$\frac{\text{MS among}}{\text{MS within}}$	
Within groups	$\sum n - a$	SS total − SS among	$\frac{\text{SS within}}{\Sigma n - a}$		
Total	$\sum n - 1$	$\sum\sum Y^2 - \frac{(\sum\sum Y)^2}{\sum n}$			

We now do an ANOVA with the data set called "number of cups of rice consumed." A cultural anthropologist is interested in studying dietary habits in a community divided into three ethnic groups. The researcher wishes to determine if the groups consume different quantities of rice, a prized food source in two of the groups. The anthropologist records the number of cups of uncooked rice prepared and consumed per week in random samples from the three groups. The households have been matched for the number of persons in the household. The necessary summary data are shown below.

$n_1 = 15$	$n_2 = 15$	$n_3 = 15$
Thus, $\sum n = 15 + 15 + 15 = 45$.		
$\sum Y_1 = 44$	$\sum Y_2 = 129$	$\sum Y_3 = 145$
Thus, $\sum\sum Y = 44 + 129 + 145 = 318$.		
$\sum Y_1^2 = 172$	$\sum Y_2^2 = 1137$	$\sum Y_2^2 = 1419$
Thus, $\sum\sum Y^2 = 172 + 1137 + 1419 = 2728$.		
$\bar{Y}_1 = 3.93$	$\bar{Y}_2 = 8.6$	$\bar{Y}_3 = 9.67$
$s_1 = 1.75$	$s_2 = 1.40$	$s_3 = 1.11$
$s_1^2 = 3.07$	$s_2^2 = 1.97$	$s_3^2 = 1.23$

We follow the usual steps when testing a hypothesis:

1. We state the null and alternative hypothesis: H0: $\mu_1 = \mu_2 = \mu_3$, H1: $\mu_1 \neq \mu_2 \neq \mu_3$.
2. The alpha level will be 0.05. The degrees of freedom are df = 2 and 42 ($\sum n - a = 45 - 3$). For these degrees of freedom the critical value is not listed in Table 6.1. Therefore, we use the more conservative values for df = 2, 40. The critical values at these degrees of freedom are 3.23 and 5.18 at 0.05 and 0.01 respectively.
3. The sample has been collected and given to us.
4. The F ratio is computed. To that end, it is best to compute first the SS total, then the SS among, and last the SS within by subtracting the SS among from the SS total. Notice that the second term of the SS total equation is used again in the SS among.

$$\text{SS total: } \sum\sum Y^2 - \frac{\left(\sum\sum Y\right)^2}{\sum n} = 2728 - \frac{(318)^2}{45} = 2728 - \frac{101{,}124}{45}$$
$$= 2728 - 2247.2 = 480.8$$

$$\text{SS among: } \sum \frac{\left(\sum Y_i\right)^2}{n_i} - \frac{\left(\sum\sum Y\right)^2}{\sum n} = \frac{(44)^2}{15} + \frac{(129)^2}{15} + \frac{(145)^2}{15} - 2247.2$$
$$= \frac{1936}{15} + \frac{16{,}641}{15} + \frac{21{,}025}{15} - 2247.2$$
$$= 2640.13 - 2247.2 = 392.93$$

$$\text{SS within: SS total} - \text{SS among} = 480.8 - 392.93 = 87.87.$$

$$\text{MS among: } \frac{\text{SS among}}{a-1} = \frac{392.93}{2} = 196.47.$$

$$\text{MS within: } \frac{\text{SS within}}{\sum n - a} = \frac{87.87}{42} = 2.09.$$

$$F \text{ ratio: } \frac{\text{MS among}}{\text{MS within}} = \frac{196.47}{2.09} = 94 \text{ with df} = a - 1 \text{ and } \sum n - a.$$

Our F ratio far exceeds the critical value at $\alpha = 0.0$ (CV $= 3.23$) and $\alpha = 0.01$ (CV $= 5.18$). We therefore reject the null hypothesis: at least one of the samples is different from the other. We display our results in the ANOVA table:

ANOVA Table

Source of variation	df	Sums of squares (SS)	Mean squares (MS)	F	p
Among groups	2	392.93	196.47	94	<0.01
Within groups	42	87.87	2.09		
Total	44	480.8			

Practice problem 6.1

An archaeologist is interested in comparing the size of ears of corn found at three archaeological sites from the same time period located in close proximity. The researcher suspects that one of the communities practiced agriculture more intensely, and could have produced larger ears of corn. The data are in a file called "ears of corn."

$n_1 = 19$	$n_2 = 19$	$n_3 = 28$	Thus, $\sum n = 66$.
$\sum Y_1 = 199$	$\sum Y_2 = 195$	$\sum Y_3 = 273$	Thus, $\sum\sum Y = 667$.
$\sum Y_1^2 = 2{,}203$	$\sum Y_2^2 = 2095$	$\sum Y_3^2 = 2799$	Thus, $\sum\sum Y^2 = 7097$.
$\bar{Y}_1 = 10.47$	$\bar{Y}_2 = 10.26$	$\bar{Y}_3 = 9.75$	
$s_1 = 2.57$	$s_2 = 2.28$	$s_3 = 2.25$	
$s_1^2 = 6.59$	$s_2^2 = 5.20$	$s_3^2 = 5.08$	

We follow the usual steps:

1. We state the null and alternative hypothesis: H0: $\mu_1 = \mu_2 = \mu_3$, H1: $\mu_1 \neq \mu_2 \neq \mu_3$.

2. The alpha level will be 0.05. The degrees of freedom are df = 2 and 62. For these degrees of freedom the critical value is not listed in Table 6.1. Therefore, we use the more conservative values for df = 2, 60.
3. The sample has been collected and given to us.
4. The F ratio is computed.

$$\text{SS total: } \sum\sum Y^2 - \frac{(\sum\sum Y)^2}{\sum n} = 7097 - \frac{(667)^2}{66} = 7097 - \frac{444{,}889}{66}$$
$$= 7097 - 6740.74 = 356.26.$$

$$\text{SS among: } \sum \frac{(\sum Y_i)^2}{n_i} - \frac{(\sum\sum Y)^2}{\sum n} = \frac{(199)^2}{19} + \frac{(195)^2}{19} + \frac{(273)^2}{28} - 6740.74$$
$$= 6747.3289 - 6740.74 = 6.59.$$

$$\text{SS within: SS total} - \text{SS among} = 356.26 - 6.59 = 349.67.$$

$$\text{MS among: } \frac{\text{SS among}}{a-1}\ \frac{6.59}{2} = 3.295.$$

$$\text{MS within: } \frac{\text{SS within}}{\sum n - a}\ \frac{349.67}{63} = 5.55.$$

$$F\text{ratio: } \frac{\text{MS among}}{\text{MS within}} = \frac{3.295}{5.55} = 0.59 \text{ with df } = a-1 \text{ and } \sum n - a.$$

Our F ratio is smaller than the critical value (2.76 at $\alpha = 0.05$). As a matter of fact, any time an F ratio is under the value of one, we know that it is not significant. For that reason, we write in the ANOVA table "ns" instead of a p value, for "not significant." Therefore, we do not reject the null hypothesis: there is no evidence that the size of ears of corn differed among the three sites. We display our results in the ANOVA table:

Source of variation	df	Sums of squares (SS)	Mean squares (MS)	F	p
Among groups	3	6.59	3.29	0.59	ns
Within groups	63	349.57	5.55		
Total	65	356.26			

6.3 ANOVA assumptions

As should be done with a t test, the assumptions of one-way ANOVA should be tested before the actual test is done. The assumptions of ANOVA remain the same as those mentioned in chapter five. Likewise, we only need to test for two assumptions:

1. *Random sampling.* The samples were collected giving each member of the population an equal chance of being selected.
2. *Independence of variates.* The collection of an observation does not influence which other observations are collected within and outside of the sample.
3. *Data are normally distributed.* This is best tested with a computer package.
4. *Homoscedasticity.* If you test this assumption by hand you would do it with the F_{max} test, dividing the largest variance by the smallest variance (as done in chapter five).

Practice problem 6.2

Let us test the assumptions of homoscedasticity and normality for the three samples in practice problem 6.1. For the homogeneity-of-variances-assumption we need the largest and smallest variances, and compute: $F_{max} = \frac{6.59}{5.08} = 1.297$. For df = 2 and 15, the critical value at 0.05 is 2.86. Therefore we do not reject the null hypothesis and conclude that the variances are homogeneous (see Table 5.1). The test for the null hypothesis that the distribution of length is normal in each sample yielded non-significant statistics for the first two samples (not shown) which led us to accept the null hypothesis of normal distribution. However, we got contradictory results for the third sample, and I am showing these below:

Name of test	Statistic	*p* value
Shapiro–Wilk	W 0.912803	0.0231
Kolmogorov–Smirnov	D 0.138322	0.1500
Cramer–von Mises	W-Sq 0.100271	0.1075
Anderson–Darling	A-Sq 0.734574	0.0491

Here we have another case of contradiction in results among the four tests for normal distribution. The studious reader might want to look into issues of kurtosis and skewness for this sample.

6.4 Post-hoc tests

If a model I ANOVA null hypothesis is rejected, we conclude that at least one of the samples did not come from the population proposed by H0. We then need to employ other tests to determine which samples differ from which. It is not appropriate to perform *t* tests repeatedly among the several possible pairs of means because the probability associated with committing a type I error in each of these tests is one out of 20 (1/20 = 0.05). If this is the risk of committing a type I error we are willing to take, then that risk should remain stable *throughout the experiment or analysis*. If we first use a 0.05 α level for the ANOVA, and proceed to compute a number of *t* tests, we increase our type I error probability for the entire experiment (indeed, if the results of these t tests are plotted, they form a normal distribution, so that 5% of them would be significant as a result of the α level we have chosen). Therefore, what we need are tests whose α level is more conservative than 0.05 so that if we reject the null hypothesis, our chance of having committed an error is still 0.05, not higher.

The tests we cover in this book are known as post-hoc, **un-planned**, or a-posteriori tests, because they compare all groups after an ANOVA is found to be significant. There are other types of tests known as a-priori or **planned** tests which are planned before the ANOVA takes place. These would be used in a laboratory setting, in which, for example, a group of animals receives a "regular" diet (the control), and several groups receive different experimental diets. If the researcher is interested in comparing the effect of the control against *all* experimental groups, then she could perform a planned comparison after the ANOVA. This experimental setting, however, is rarely desirable in anthropology, where we are more interested in comparing all groups with each other, should the ANOVA be significant. Therefore, only post-hoc tests will be covered here. There are a large number of tests available (least significant difference (LSD), the GT2, the T', the Tukey–Kramer method, the Welsch step-up procedure, the Scheffé test, etc.), and PASW and SAS provide ready access to them. Therefore, it is unnecessary to cover the mathematical computations of more than one such test.

6.4.1 The Scheffé test

This test was chosen over the others because it is known to be the post-hoc test which is the most conservative. That is, it is the test which is least likely to produce type I errors. The test proceeds as follows: for each comparison between two group means, a **SS between** is computed and is divided by the df among (from the ANOVA table) to obtain a MS between. An F ratio is obtained by dividing the MS between by the MS within from the ANOVA table. The F ratio tests the null hypothesis that the two samples were obtained from the same population, and has df $= a - 1$ and $\sum n - a$. Only the SS between needs to be computed per comparison. Everything else (the MS within and the degrees of freedom) is taken from the ANOVA table which has been already computed. The formula for the SS between is shown in formula 6.2.

Formula 6.2

Formula for the SS between for the Scheffé test.

$$\text{SS between} = \sum \frac{(\sum Y_i)^2}{n_i} - \frac{(\sum \sum Y)^2}{\sum n}$$

where only the data of the two groups are considered.

The data first discussed (cups of rice used by three different communities) are used to illustrate the Scheffé test. The following information is necessary:

$n_1 = 15$	$n_2 = 15$	$n_3 = 15.$
$\sum Y_1 = 44$	$\sum Y_2 = 129$	$\sum Y_3 = 145$
$\sum Y_1^2 = 172$	$\sum Y_2^2 = 1137$	$\sum Y_2^2 = 1419$
$\bar{Y}_1 = 3.93$	$\bar{Y}_2 = 8.6$	$\bar{Y}_3 = 9.67$
$s_1 = 1.75$	$s_2 = 1.40$	$s_3 = 1.11$
$s_1^2 = 3.07$	$s_2^2 = 1.97$	$s_3^2 = 1.23$

The F ratio is compared with the same critical value used previously, since the degrees of freedom are the same: df = 2, 40. All SS between will be divided by 2, the df among, and all MS between by 2.09, the MS within.

Comparison of groups 1 and 2. Thus: $\sum\sum Y = 44 + 129 = 173$, and $\sum n = 15 + 15 = 30$.

SS between groups 1 and 2:

$$= \left[\frac{(44)^2}{15} + \frac{(129)^2}{15}\right] - \frac{(173)^2}{30} = 1238.47 - 997.63 = 240.84$$

$$\text{MS between} = \frac{240.84}{2} = 120.42$$

$$F = \frac{\text{MS between}}{\text{MS within}} = \frac{120.42}{2.09} = 57.62.$$

The F ratio far exceeds the critical values shown in Table 6.1 and even the critical value at $\alpha = 0.001$ (CV = 8.25; taken from the complete F table). Thus, the null hypothesis that groups 1 and 2 were obtained from the same population is rejected.

Comparison of groups 1 and 3. Thus: $\sum\sum Y = 44 + 129 = 173$, and $\sum n = 15 + 15 = 30$.

SS between groups 1 and 3:

$$= \left[\frac{(44)^2}{15} + \frac{(145)^2}{15}\right] - \frac{(189)^2}{30} = 1530.73 - 1190.7 = 340.03$$

$$\text{MS between} = \frac{340.03}{2} = 170.015$$

$$F = \frac{170.015}{2.09} = 81.35.$$

Once again, the null hypothesis is rejected with great confidence: groups 1 and 3 were not obtained from the same population.

Comparison of groups 2 and 3. Thus: $\sum\sum Y = 129 + 145 = 274$, and $\sum n = 15 + 15 = 30$.

SS between groups 2 and 3:

$$= \left[\frac{(129)^2}{15} + \frac{(145)^2}{15}\right] - \frac{(274)^2}{30} = 2511.07 - 2502.53 = 8.53$$

$$\text{MS between} = \frac{8.53}{2} = 4.27$$

$$F = \frac{4.27}{2.09} = 2.04.$$

The F ratio is smaller than the critical value at $\alpha = 0.05$ (CV = 3.23). Thus, the null hypothesis is not rejected. The general conclusion of this study is that group 1 eats a significantly different number of cups of rice per week from groups 2 and 3. The latter two groups do not differ in their consumption. By the time you would have done this ANOVA you would have already computed descriptive statistics and done a few graphs, so perhaps you would have expected these results.

6.5 Model I, two-way ANOVA

We might be interested in determining whether the weight of children is affected by SES and do a one-way ANOVA with three groups of children classified by their SES. We might also be interested in whether the weight of children is affected by gender and do another one-way ANOVA (in the form of a t test) with children classified by their gender. A two-way ANOVA design allows us to look at the two treatment effects (say, SES and gender) *and the possible interaction of these two treatment effects* at the same time. Because both independent variables are seen as treatment effects, this is a model I, two-way ANOVA. The advantages of doing one two-way ANOVA is that we decrease our probability of a type I error (because we do only one ANOVA instead of two), and that we save time and even money (because we work with fewer subjects) than if we do two one-way ANOVAs.

Let us think again of ANOVA as **a form of regression analysis**, in which we seek to explain the dependent variable as follows: $\hat{Y}_{ijk} = \mu + \alpha_i + \beta_j + (\alpha\beta)_{ij} + \epsilon_{ijk}$ where $\hat{Y}_{ij}$ = the predicted y for the kth observation who is in the subgroup defined by the α_i treatment effect associated with the first independent variable and by the β_j treatment effect associated with the second independent variable, μ = the mean of all observations in the study regardless of the group to which they belong, $(\alpha\beta)_{ij}$ is the interaction between the treatment effects, and ϵ_{ijk} is the normal random variation found within each group which is measured in the kth observation. The treatment effects are denoted with the symbols α and β.

Interaction is a tricky concept which might make more sense with a clinical example. It is well known that if a child suffers from mild malnutrition and a mild infection he will be much more ill than if he only had one of the two ailments. This is because both conditions have a synergistic effect on the child's health, i.e. when both are present they make each other much worse. A **synergistic effect** between alcohol and numerous drugs (from anti allergic to muscle relaxing drugs) is what puts patients to sleep if they are not cautious and take the drug with alcohol. In sum **synergy** between two treatment effects enhances their individual effects. The other way in which interaction between two treatment effects may occur is in the form of **interference**, in which the presence of both effects inhibits their action. For example, when a drug is taken in the presence of a nutrient, the drug's effects might be nullified. Thus, subjects who take the drug without the nutrient will be affected by the drug while subjects who take the drug *and* the nutrient will not be affected by the drug because the nutrient interferes with the drug (you may have heard that high blood pressure patients under therapy with some drugs may not consume grapefruit).

In two-way ANOVAs *we test more than one null hypothesis*, namely, we are asking: (1) Does the model explain a significant portion of the variation of the Y? By model we mean the two Xs and their interaction. (2) Does the first independent variable affect the dependent variable? (For example, do children of different SES vary in their body weight?) (3) Does the second independent variable affect the dependent variable? (For example, do boys and girls differ in their body weight?) (4) Is there interaction between the two treatment effects? (Are boys heavier than girls in two out of three categories of SES *but* lighter than girls in the third category?) If there is no interaction we can produce "nice and clean" conclusions. For example, SES does not affect body weight although gender does: boys are heavier than girls. However, if there is interaction we are unable to make such "nice and clean" statements because boys are heavier than girls only in some SES categories but not in others. Therefore the first thing we need to do is to test for the presence of interaction. If there is interaction then we cannot make any statements about how the independent variables affect the dependent variable with a two-way ANOVA. If there is interaction, the researcher would need to analyze the data with two separate one-way ANOVAs.

I mentioned that doing one two-way ANOVA instead of two one-way ANOVAs may save us money because we can use the same subjects to test for the presence of two treatment effects in one study. This is true, although most two-way ANOVAs deal with few subjects per subgroup (defined by the two independent variables). Indeed, sample size becomes a challenge in many two-way ANOVAs. For this reason the testing of normality or homoscedasticy assumptions becomes a futile effort in most cases. Even if the researcher is working with subjects that are much easier to manipulate than humans, it is still a challenge to get enough (say) rats who are male and female and who are exposed to each of (say) three levels of an independent variable.

The formulae for computing a two-way ANOVA with equal sample sizes are shown below. The hand computations for unequal sample sizes are quite complex. Fortunately both SAS and PASW easily handle unequal size situations. The formulae may look overwhelming at first but they are all based on the same sums of squares formulae you know so well.

We first compute the total sums of squares (the total pie of variation, i.e. SS_Y), which we split into two large pieces: the model sums of squares and the within or error sums of squares. The former quantifies the amount of variation of the *Y* explained by the model, where the model includes the two independent variables and their interaction. With the model SS, we compute a MS (by dividing the SS by its df, as is usual).

We use the model MS to test the null hypothesis that the model does not explain a significant portion of the variation of the *Y* (here we are making direct reference to the fact that ANOVA is a form of regression analysis, and that our model includes two independent variables and their interaction). If we accept the H0 we accept the proposal (for example) that neither SES nor gender nor their interaction affect weight. However, if we reject this hypothesis we know that at least one of the treatment effects (or their interaction) did have a significant effect on weight. Therefore we need to partition the SS model into three pieces: (1) The treatment effect due to one of the independent variables (let us say that this is the one that we will put in the rows of the data set, so let us call it row SS); (2) The treatment effect due to the other independent variable (let us say that this is the one that we will put in the columns of the data set, so let us call it column SS); and (3) The interaction between both treatment effects. The computation of the two-way ANOVA is explained in formula 6.3 below.

Formula 6.3

Formula for the computation of two-way ANOVA.
Equal sample sizes

Source of Variation	df	SS	MS
Model	$rc - 1$	$\sum^{r}\sum^{c}\frac{(\sum^{n} Y)^2}{n} - \frac{(\sum^{r}\sum^{c}\sum^{n} Y)^2}{rcn}$	$\frac{SS\ model}{rc - 1}$
Among rows	$r - 1$	$\frac{\sum^{r}(\sum^{c}\sum^{n} Y)^2}{cn} - \frac{(\sum^{r}\sum^{c}\sum^{n} Y)^2}{rcn}$	$\frac{SS\ rows}{r - 1}$
Among columns	$c - 1$	$\frac{\sum^{c}(\sum^{r}\sum^{n} Y)^2}{rn} - \frac{(\sum^{r}\sum^{c}\sum^{n} Y)^2}{rcn}$	$\frac{SS\ columns}{c - 1}$
Interaction	$(r - 1)(c - 1)$	SS model – SS rows – SS columns	$\frac{SS\ interaction}{(r - 1)*(c - 1)}$
Error	$rc(n - 1)$	SS total – SS model	$\frac{SS\ error}{rc(n - 1)}$
Total	$rcn - 1$	$\sum^{r}\sum^{c}\sum^{n} Y^2 - \frac{(\sum^{r}\sum^{c}\sum^{n} Y)^2}{rcn}$	

Let us discuss these formulae before we apply them.

$$\text{The SS total is: } \sum^{r}\sum^{c}\sum^{n} Y^2 - \frac{(\sum^{r}\sum^{c}\sum^{n} Y)^2}{rcn}$$

The second element on the total SS will be used in the computation of several of the sums of squares and it is frequently called the correction term (CT). To compute it you add all of the observations to obtain the grand total, square this number, and divide by the product of the sample size*the number of rows*the number of columns. The element on the left is the grand total of squared observations. You can see how similar this formula is to the sums of squares we have computed before.

$$\text{The SS model is: } \sum^{r}\sum^{c} \frac{(\sum^{n} Y)^2}{n} - \frac{(\sum^{r}\sum^{c}\sum^{n} Y)^2}{rcn}$$

The element on the right is the CT. The element on the left is the sum of the total of the observations in each of the subgroups defined by the two variables (for example, the group of males in SES 1, the group of females in SES 2, etc.) squared divided by sample size. The formula is instructing you to take the sum of each group across the columns and across the rows.

$$\text{The SS rows is: } \frac{\sum^{r}(\sum^{c}\sum^{n} Y)^2}{cn} - \frac{(\sum^{r}\sum^{c}\sum^{n} Y)^2}{rcn}$$

The element on the right is the CT. The element on the left is the total of the observations of row one squared plus the total of row two squared and so forth divided by the number of columns*sample size.

$$\text{The SS column is: } \frac{\sum^{c}(\sum^{r}\sum^{n} Y)^2}{rn} - \frac{(\sum^{r}\sum^{c}\sum^{n} Y)^2}{rcn}$$

The element on the right is the CT. The element on the left is the total of the observations of column one squared plus the total of column two squared, and so forth divided by the number of rows*sample size.

The SS interaction is obtained by subtraction.

The SS error is obtained by subtraction.

MS computations:	
MS rows:	SS rows/row df
MS column:	SS column/column df
MS error:	SS error/error df
MS interaction:	SS interaction/interaction df
Hypothesis testing:	Divide the appropriate mean square by the MS error

Practice problem 6.3

An anthropologist is interested in determining whether children of both genders get the same number of years of formal education in three different religious groups in a village. The anthropologist obtained 20 subjects of age 30 of both genders in each of the three ethnic communities, and he recorded the number of years of formal education completed by them. The dependent variable is $Y =$ years of formal education, and the independent variables are $X_1 =$ gender (where there are two rows so $r = 2$) and $X_2 =$ religion (where there are three columns, so $c = 3$). Do groups of individuals defined by religion and gender differ in their years of formal education? (This is going to be tested by the model MS.) If the answer is yes, that is, if we reject the model null hypothesis, we can ask: (1) if members of both genders differ (this is going to be tested by the row MS), (2) if members of the three religious communities differ (this is going to be tested by the column MS), and (3) if there is an interaction between religion and gender (this is going to be tested by the interaction MS). I am reproducing the data set here because it is important that you see what I mean by rows (the two genders) and by columns (the three religions). The data set is found in Excel format at the book's website and it is called "Education in children of two genders and three religions." Please note that there are six subgroups: religion one and gender one, religion one and gender two, . . . , religion three and gender one, and religion three and gender two.

	Religion			
	One	Two	Three	
Gender				
1	12	9	14	
1	10	8	15	
1	9	13	15	
1	13	11	16	
1	8	13	16	
1	9	11	14	
1	12	15	14	
1	11	12	19	
1	12	14	13	
1	12	10	13	
1	12	12	15	
1	9	12	14	
1	12	10	17	
1	11	15	20	
1	14	7	16	
1	7	15	13	
1	11	10	16	
1	11	12	15	
1	14	12	16	
1	8	11	13	
Sum	**217**	**232**	**304**	**753**

(cont.)

		Religion			
		One	Two	Three	
	Gender				
	2	14	12	9	
	2	15	14	9	
	2	14	15	12	
	2	14	15	8	
	2	13	13	8	
	2	10	13	10	
	2	13	15	7	
	2	14	16	12	
	2	15	13	8	
	2	15	18	12	
	2	16	15	10	
	2	14	17	10	
	2	18	15	7	
	2	12	11	8	
	2	18	16	9	
	2	13	14	13	
	2	15	12	14	
	2	14	13	9	
	2	12	11	12	
	2	13	14	10	
	Sum	**282**	**282**	**197**	**761**
Column sums		**217 + 282 = 499**	**232 + 282 = 514**	**304 + 197 = 501**	
Total sum					**1514**

We follow the usual steps:

1. We state the null and alternative hypothesis: H0: $\mu_1 = \mu_2 \ldots = \mu_6$. H1: At least one of the subgroups differs. The degrees of freedom to test this H0 are: $rc - 1$, $rc(n-1) = 2^*3 - 1, 2^*3^*19 = 5, 114$. Since we do not have exactly these degrees of freedom in Table 6.1 we will use the critical value at the more conservative degrees of freedom 5, 100 (critical value = 2.30 at $\alpha = 0.05$, or 3.20 at $\alpha = 0.01$). If we reject this null hypothesis then we will test the following three hypothesis:
2. H0: Interaction of religion and gender = 0. The degrees of freedom to test this H0 are $r - 1^*c - 1, rc(n-1) = 2 - 1, 2^*3^*19 = 2, 114$. Since we do not have exactly these degrees of freedom in Table 6.1 we will use the critical value at the more conservative degrees of freedom 2, 100 (CV = 3.09 at $\alpha = 0.05$, or 4.82 at $\alpha = 0.01$). If we do not reject this H0 we test the following two hypotheses.
3. H0: $\mu_{gender\,1} = \mu_{gender\,2}$. The degrees of freedom to test this H0 are $r - 1, rc(n - 1) = 2 - 1, 2^*3^*19 = 1, 114$. Since we do not have exactly these degrees of freedom in Table 6.1 we will use the critical value at the more conservative degrees of freedom 1, 100 (cv = 3.94 at $\alpha = 0.05$ or 6.90 or $\alpha = 0.01$).

4. H0: $\mu_{religion\,1} = \mu_{religion\,2} = \mu_{religion\,3}$. The degrees of freedom to test this H0 are $c - 1$, $rc(n - 1) = 3 - 1$, $2*3*19 = 2{,}114$. Since we do not have exactly these degrees of freedom in Table 6.1 we will use the critical value at the more conservative degrees of freedom 2, 100 (CV = 3.09 at $\alpha = 0.05$, or 4.82 at $\alpha = 0.01$).

It is easier to compute the necessary quantities for the computation of the sums of squares. Thus: $\left(\sum^{r}\sum^{c}\sum^{n} Y\right) = 217 + 232 + 304 + 282 + 197 = 1514$ This is the grand total (where 217, 232 and so forth are the ΣY for each group). $\sum^{r}\sum^{c}\sum^{n} Y^2 = 12^2 + 10^2 + \ldots + 12^2 + 10^2 = 2429 + 4044 + 2786 + 4044 + 4690 + 2019 = 20{,}012$. This is the grand total of squared observations, where 2429, 4044, and so forth are the $\sum Y^2$ for each group.

$$\text{SS total: } \sum^{r}\sum^{c}\sum^{n} Y^2 - \frac{\left(\sum^{r}\sum^{c}\sum^{n} Y\right)^2}{rcn}$$

$$\text{SS total: } 20{,}012 - \frac{(1514)^2}{2*3*20} = 20{,}012 - 19{,}101.6333 = 910.3667.$$

$$\text{SS model: } \sum^{r}\sum^{c} \frac{\left(\sum^{n} Y\right)^2}{n} - \frac{\left(\sum^{r}\sum^{c}\sum^{n} Y\right)^2}{rcn}$$

$$= \frac{217^2}{20} + \frac{232^2}{20} + \frac{304^2}{20} + \frac{282^2}{20} + \frac{197^2}{20} = 19{,}559.3 - 19{,}101.6333 = 457.6667.$$

$$\text{MS model} = \frac{457.6667}{5} = 91.53334.$$

$$\text{SS error: SS total} - \text{SS model} = 910.3667 - 457.667 = 452.7.$$

$$\text{MS error: } \frac{452.7}{114} = 3.97105.$$

F ratio for H0 $\mu_1 = \mu_2 \ldots = \mu_6$. H1: At least one of the groups differs: $F = \frac{91.53334}{3.97105} = 23.05$. Our F ratio far exceeds the critical values at df = 5, 100 [2.30 ($\alpha = 0.05$), or 3.20 ($\alpha = 0.01$)]. We reject the null hypothesis, and conclude that at least one of the groups differs. Now we need to further cut the slice of the pie explained by the model, and determine if there is interaction. If there is no interaction we can test if the groups differ by religion or by gender or by both. We compute the SS interaction and the MS interaction. We divide the MS interaction by MS within (already computed) to obtain the F ratio to test this H0.

SS interaction = SS model – SS rows – SS columns. Therefore we first compute the SS rows and SS columns. We will subtract them from the SS model, which is already computed.

$$\text{SS rows: } \frac{\sum^{r}\left(\sum^{c}\sum^{n} Y\right)^2}{cn} - \frac{\left(\sum^{r}\sum^{c}\sum^{n} Y\right)^2}{rcn}$$

Table 6.3 ANOVA table for practice problem 6.3.

Source	df	Sum of squares	Mean square	*F* value	*p*
Model	5	457.6667	91.53334	23.05	<0.0001
religion	2	3.3167	1.65835		
gender	1	0.53337	0.53337		
gender*religion	2	453.81663	226.908315	57.14	<0.0001
Error	114	452.70	3.97105		
Total	119	910.3667			

SS rows $= \frac{753^2+761^2}{3*20} - 19{,}101.6333 = 19{,}102.16667 - 19{,}101.6333 = 0.53337$. No need to compute the MS rows just yet because we do not know if we will use it to test the hypothesis that different genders differ in their education. We need to wait to see if there is interaction.

$$\text{SS columns: } \frac{\sum^{c}\left(\sum^{r}\sum^{n} Y\right)^2}{rn} - \frac{\left(\sum^{r}\sum^{c}\sum^{n} Y\right)^2}{rcn}$$

$$\begin{aligned}\text{SS columns} &= \frac{499^2 + 514^2 + 501^2}{2*20} - 19{,}101.6333 \\ &= 19{,}104.95 - 19{,}101.6333 = 3.3167.\end{aligned}$$

No need to compute the MS columns just yet because we do not know if we will use it to test the hypothesis that different religions differ in their education. We need to wait to see if there is interaction.

$$\begin{aligned}\text{SS interaction} &= \text{SS total} - \text{SS row} - \text{SS column} = 457.6667 - 0.53337 - 3.3167 \\ &= 453.81663. \\ \text{MS interaction} &= \frac{453.81663}{2} = 226.9083.\end{aligned}$$

F ratio for H0: Interaction of religion and gender = 0: $\frac{226.9083}{3.9710} = 57.14$. This *F* ratio far exceeds the critical values at degrees of freedom 2, 100: 3.09 ($\alpha = 0.05$), or 4.82 ($\alpha = 0.01$). We reject the null hypothesis of interaction with great confidence.

The results of our ANOVA are shown in Table 6.3, which for completeness has the MS for rows and columns computed. Our analysis cannot go any further: we have shown that there is interaction between the two independent variables, so we cannot make a general statement about how religion and gender affect the number of years of formal education because there is not a directional manner in which the independent variables affect the dependent variable. Perhaps a graph will help us understand the interaction between religion and gender. Figure 6.5 shows the mean number of years of education for both genders by religion. The graph shows that in religious groups one and two, gender two receives more years of education than gender one. However, in religious group three, gender two receives fewer years of education than gender one. Therefore

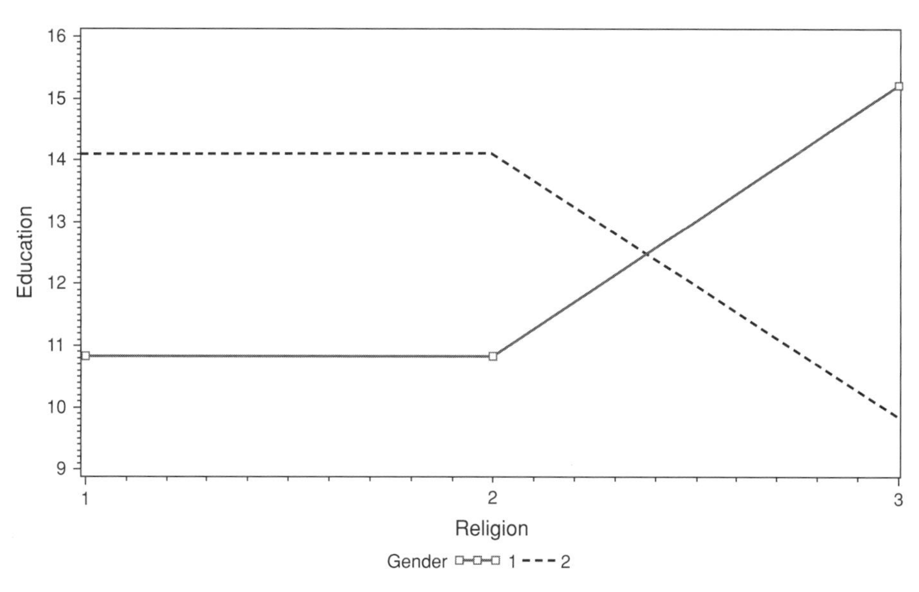

Figure 6.5 Interaction between gender and religion.

in our data set we cannot make any general statements about how gender and religious affiliation affect the years of education received. We would need to follow up this study with two one-way ANOVAs.

6.6 Other ANOVA designs

We have discussed a model I, one-way ANOVA in which groups defined by one independent variable are subjected to treatment effects. We have also discussed a model I, two-way ANOVA in which groups defined by two independent variables are subjected to two treatment effects. It is also possible to have a **mixed-model two-way ANOVA**, in which one of the independent variables follows a model I (a repeatable treatment) while the second one follows a model II design (it is not a repeatable treatment). You will recall that a model II ANOVA design divides the subjects according to un-repeatable, random effects. I have to credit my younger daughter with giving me an example for a mixed-model ANOVA when she was thinking of her science project. My daughter is disturbed by the fact that apples change their color after they are sliced. As she was eating apple slices quickly to avoid eating them after they changed their color, she wondered if the type of apple affects the time it takes for the slices to change color. This is a model I question: we can divide the apples according to variety, and time how long it takes for equally sliced pieces to change color. Let us say that we will have three rows with three varieties: Granny Smith, Red Delicious, and Gala. The question we would also like to know is: Is there significant variation among different apples of the same variety? This is something that is completely out of our power to manipulate. It is entirely a matter of chance that I picked three random apples within any of these varieties. Therefore, this second level of classification is a model II ANOVA, with which I will be able to determine if there is a significant added portion of variance due to random factors within an apple variety. One good reason to do this as a pilot study is that if we determine that apples within variety do not differ significantly then we will be content to slice only one apple instead of ten, and we will use our resources wisely (before you accuse me of being an overpowering mother, be assured that I told my daughter that I thought her idea was good, and said nothing else!).

ANOVA in anthropology

Linear enamel hypoplasia is a well-known marker of stress during the period of growth and development. Bioarchaeologists wish to estimate the probable age at which stress occurred during the formation of the tooth to be able to make statements about when children are most vulnerable in a particular group. The problem of course is how to determine the age of the insult from the teeth. Martin *et al.* (2008) compare two methods for estimating the age at which the insult occurred in the anterior dentition in a set of teeth from the Hammann–Todd collection. A mixed model was most appropriate here because while the two different methods of estimating age in the tooth are repeatable treatments, the individual tooth on which the measures were taken is not (Martin *et al.* 2008).

Is the present-day ecology of paleontological sites similar to the ecology of the site when fossils were being formed? Has the environment changed? In an attempt to answer these questions about the site where the Taung child fossil was found, Williams and Patterson compare the number of wear scars in teeth of living and extinct papionins from Taung. The authors do a one-way ANOVA where the groups are defined by the species. They compare the number of fine scratches, coarse scratches, large pits, and small pits and reject the null hypothesis of equality for all of these features. The authors then follow up the ANOVA with a post-hoc test in which each species is compared with the others. Williams and Patterson conclude that in comparison with present-day South Africa, Taung had a wetter environment (Williams and Patterson 2010).

6.7 Chapter 6 key concepts

- What is the difference between model I and model II?
- What is the difference between one-way and two-way ANOVA?
- Explain mixed-model ANOVA.
- When do you use post-hoc tests? Why don't we compare groups with *t* tests?
- Assumptions of ANOVA.
- What should you do if your data violate the assumptions?

6.8 Computer resources

1. PASW allows you to request a homoscedasticity test in the following menu: Analyze → compare means → ANOVA → options. As far as I can tell SAS does not allow you to request this as part of its menu ANOVA options.

6.9 Chapter 6 exercises

All data sets are available for download at www.cambridge.org/9780521147088.

1. Write down a research project in your field of study in which you might use a model I one-way ANOVA, and a model I two-way ANOVA. Is there any chance you might use a mixed-model ANOVA?
2. Use the data set called "Rangia shells in tree levels." Do an ANOVA testing the null hypothesis that mean weight (measured in grams) differs between the three levels. Test the assumptions.
3. Use the data set called "Birth weights in three groups" to do a two-way ANOVA in which the dependent variable is weight (rounded to the nearest pound) and the two independent variables are baby's sex and baby's ethnicity.

7 Non-parametric tests for the comparison of samples

Testing the assumptions of parametric tests should be a life-long practice. An analyst has two broad options if parametric test assumptions are not met: (1) The data can be transformed so that they have a normal distribution and may be analyzed with parametric tests. Transformations will be discussed in the chapter on simple linear regression since transformations are more frequently performed when doing a regression analysis. (2) The data may be analyzed with non-parametric tests which allow the researcher to answer the original question with a test that does not have the assumptions of a parametric test.

Non-parametric tests are so called because they make no reference to the parameters of the population from which the sample was obtained. Indeed, we are going to start phrasing our null hypotheses with words instead of parametric notation. These tests are also called distribution-free tests because we make no assumptions about the shape of the distribution of the population from which our sample was taken. Non-parametric tests can also be used when our sample sizes are small (say under 20), at which point we may not even want to test for normality in a sample. Anthropologists are frequently forced to work with exceedingly small samples because of the nature of our data (number of fossils, or number of non-human primate childbirths observed in the wild, etc.) so these tests have a broad appeal in the field. Lastly, non-parametric tests can be used with data that are not precisely measured. For example, if you are taking a survey on the number of domestic animals living in each household you may want to have categories such as 0, 1–2, 3–4 . . . 9–10, more than 10. The latter category is not precise since it contains everything from 11 to who-knows-what. However, non-parametric tests allow us to incorporate these types of data into a statistical analysis permitting us to test hypotheses about the sample.

If we can answer the same question with a parametric test *and* avoid testing those pesky assumptions, why not go for the path of least resistance and use non-parametric tests at once? The reason parametric tests are favored over non-parametric tests is that the former are more powerful than the latter. Therefore, if the treatment effect *does* affect the dependent variable but it does so in a subtle manner, we might not be able to detect the treatment effect with a non-parametric test while we might be able to with a parametric one. Thus, all other things being equal, we should apply a parametric test.

Each analyst needs to come to a decision on which test to use on his or her own, considering the pros and cons of all options. My advice is the following: If your sample size is very small (fewer than 20) do not attempt to do a parametric test. If your sample size is adequate (over 30) and your data violate the assumptions of parametric tests, try to transform the data so that you can still use a parametric test. If your sample is between 20 and 30 perform non-parametric tests because they are easier to do than are transformations.

Just one more "warning" before we embark in the non-parametric boat: it is my experience that the algorithm used by both PASW and SAS for the computation of some of these tests is different from the formulae presented in all textbooks I have read on non-parametric tests. Therefore, it has happened to me with some frequency that the value of the statistics I compute is different from that computed by either PASW or SAS although the statistical conclusion (failure to reject or rejection of the H0) is the same. Another minor issue is the name of the statistical tests we will cover here. The test better known as the Mann–Whitney U test is a modification of an earlier statistical test advanced by Wilcoxon. For this reason in SAS the Mann–Whitney U test is referred to as the Wilcoxon (ranked data) test. In contrast, Wilcoxon developed a test for which he is better known which is called the Wilcoxon test, the Wilcoxon signed, or the Wilcoxon signed-rank test for a paired design. You just need to be careful that you know what you are requesting your computer to do.

This chapter covers three non-parametric tests, to be used instead of an un-paired *t* test (the Mann–Whitney U test), a one-way model I ANOVA (the Kruskal–Wallis test), and a paired *t* test (the Wilcoxon signed-rank test). These tests have a procedure in common: they rank the data before the statistics are computed. Thus, we first learn how to rank data.

7.1 Ranking data

The procedure for ranking data in the non-parametric tests covered in this chapter is basically the same, except that two of the tests consider the sign of the digit, whereas the other one works with the absolute value of the number. Let us practice both kinds of ranking with the following numbers: −2, 6, 1. If we rank considering the sign, then the ranking (from lower to higher) should yield:

Number	Rank
−2	1
1	2
6	3

That is, if we consider the sign, then −2 is less than 1. However, if we do not consider the sign, we need to rank the data according to their absolute value. Therefore, our ranking results should be:

Number	Absolute value	Rank
1	1	1
−2	2	2
6	6	3

A commonly confronted situation when ranking is the presence of ties. When observations are tied, the average of their rank is assigned to all of them. Therefore, we first assign ranks as if the variates were not tied, then we average these ties. For example, given the following data, we take these steps: (1) We first order the variates from lowest to highest (using the absolute value). (2) We then assign the initial ranks. (3) If some numbers are tied, we compute the average of their initial ranks. This average then becomes the final rank of the original number.

Practice problem 7.1

Ranking a set of data with ties.

Ordered raw data	Initial rank	Average ranks of tied variates	Final rank
0	1		1
1	2		2
3	**3**	$\frac{3+4}{2} = 3.5$	3.5
3	**4**		3.5
4	5		5
6	6		6
7	7		7
9	8		9
9	**9**	$\frac{8+9+10}{3} = 9$	9
9	10		9

7.2 The Mann–Whitney U test for a two-sample un-matched design

This test can be used instead of an un-paired *t* test if for any of the reasons discussed above the latter is not appropriate. This test only makes two assumptions: (1) The underlying level of measurement is continuous, although the units of observations are discrete.

Table 7.1 Selected critical values of U, the Mann–Whitney statistic for a two-tailed test at $\alpha = 0.05$.

$n_2 \backslash n_1$	5	6	7	8	9	10	11
5	2	3	5	6	7	8	9
6	3	5	6	8	10	11	13
7	5	6	8	10	12	14	16
8	6	8	10	13	15	17	19
9	7	10	12	15	17	20	23
10	8	11	14	17	20	23	26

Therefore, ties in ranking occur because of error of measurement. The researcher's only concern about this assumption is if the data contain too many ties. (2) The samples are independent. Thus, this test is not appropriate for a paired design.

The null hypothesis is that the population from which the samples are obtained is the same. If the null hypothesis is true, and the two samples were ranked together, the observations of both samples would be expected to be equally mixed. If, however, the samples came from two different populations, they would not be expected to mix, but to be on opposite sides of the ranked column of data. The test determines how evenly mixed the scores are from both samples. The Mann–Whitney U test can be applied to samples of equal or unequal sizes.

The test proceeds as follows:

1. The observations of both samples are ranked together. Ranking is done taking into consideration the sign of the observation.
2. The sum of the ranks $\sum R_1$ for the first group is computed.
3. The U statistic for both groups is computed as shown in formula 7.1.

Formula 7.1

Formula for the U statistic.

$$U_1 = (n_1)(n_2) + \frac{(n_1)(n_1 + 1)}{2} - \sum R_1, \text{ and } U_2 = (n_1)(n_2) - U_1$$

4. The *lesser* value of U, either U_1 or U_2 is chosen for significance testing. Table 7.1 shows selected critical values of U. This table requires *a different decision rule*: if the test statistic at the appropriate sample size *is less than or equal* the critical value at $\alpha = 0.05$, the null hypothesis is rejected. This table can be used if n_1 and n_2 are ≤ 20. For larger sample sizes, the U statistic can be transformed into a t or z score, and a t or normal table used (see Sokal and Rohlf 1995).

The reason for the different decision rule used with this table will be explained with two examples. In the first, the samples have ranks that do not mix at all, and in the second, the samples are ranked completely mixed. We follow the steps outlined above, starting with the first case.

1. The observations of both samples are ranked together.

Observation	Group membership	Rank
1	1	1
2	1	2
3	1	3
4	1	4
5	1	5
6	1	6
7	1	7
8	1	8
9	2	9
10	2	10
11	2	11
12	2	12
13	2	13
14	2	14
15	2	15

2. The sum of the ranks $\sum R_1$ for the first group is computed:

$$\sum R_1 = 1 + 2 + 3 + 4 + 5 + 6 + 7 + 8 = 36.$$

3. The U statistic for samples 1 and 2 (where $n_1 = 8, n_2 = 7$) is computed:

$$U_1 = (8)(7) + \frac{(8)(8+1)}{2} - 36 = 56 + 36 - 36 = 56 \text{ and}$$
$$U_2 = (8)(7) - 56 = 56 - 56 = 0.$$

4. The *lesser* value of U, either U_1 or U_2, is chosen for significance testing. The critical value at $\alpha = 0.05$ for a two-tailed test where $n_1 = 8, n_2 = 7$ is 10. Since our U is less than the critical value we reject the null hypothesis that the two samples were obtained from the same population.

In general, it is known that $U_1 + U_2 = (n_1)(n_2)$. In this example, for instance, $(n_1)(n_2) = (8)(7) = 56$, and $U_1 + U_2 = 56 + 0 = 56$. In an extreme case in which there is no overlap between the two samples, the obtained U used for hypothesis testing is 0.

Now the second example is presented. Here, the two samples are mixed:

Observation	Group membership	Rank
1	1	1
2	2	2
3	1	3
4	2	4
5	1	5
6	2	6
7	1	7
8	2	8
9	1	9
10	2	10
11	1	11
12	2	12
13	1	13
14	2	14
15	1	15

2. The sum of the ranks $\sum R_1$ for the first group is computed:

$$\sum R_1 = 1 + 3 + 5 + 7 + 9 + 11 + 13 + 15 = 64.$$

3. The U statistic for samples 1 and 2 are computed (where $n_1 = 8$, $n_2 = 7$):

$$U_1 = (8)(7) + \frac{(8)(8+1)}{2} - 64 = 56 + 36 - 64 = 28, \text{ and}$$
$$U_2 = (8)(7) - 28 = 56 - 28 = 28.$$

4. Choose the *lesser* value of U, either U_1 or U_2 for significance testing. The critical value at $\alpha = 0.05$ for a two-tailed test where $n_1 = 8$, $n_2 = 7$ is 10. In this case, either U may be used for hypothesis testing since they have the same value. The null hypothesis is accepted because the test statistic is greater than the critical value.

The reader should note that $(n_1)(n_2) = (8)(7) = 56$ and that $U_1 + U_2 = 28 + 28 = 56$. In this extreme example in which the variates were mixed by alternating group membership, the two U values were $\frac{(n_1)(n_2)}{2}$. This is the highest possible value the U can take, if the two groups do not differ. This is the reason why the null hypothesis in a Mann–Whitney U test is tested by determining how small the smallest U is. The closer it is to zero, the more likely the hypothesis is to be rejected. The test is now illustrated with a more realistic example.

Practice problem 7.2

Assume that an anthropologist is investigating the economic role of young children in two ethnically distinct groups in a community. The researcher is investigating if the children who stay home with their mothers instead of collecting food items from the forest on non-school days are of different ages. Do the two ethnic groups have different economic expectations of children when the latter are not at school? Does one group expect its young children to collect food items while the other group allows them to stay home and not contribute to the family diet? A problem the anthropologist encounters is that there are no records to validate the children's age, which must be estimated by the researcher in consultation with the mothers. To minimize error in the estimation, the researcher limits her study to one gender of children only. A non-parametric test should be used in this case because the sample sizes are small and the data are estimates, not true measures of the actual age of the children. The null hypothesis is that there is no difference in children's age by group. The data are called "children's economic contribution."

Age	Group	Rank
4	1	8.0
3	1	2.5
4	1	8.0
3	1	2.5
4	1	8.0
4	1	8.0
4	1	8.0
5	2	13.0
5	2	13.0
4	2	8.0
6	2	15.5
3	2	2.5
4	2	8.0
3	2	2.5
5	2	13.0
6	2	15.5

1. Compute the sum of the ranks for the first group:

$$\sum R_1 = 8 + 2.5 + \ldots 8 + 8 = 45.$$

2. Compute the U statistic for both groups, where $n_1 = 7$ and $n_2 = 9$:

$$U_1 = (7)(9) + \frac{(7)(7+1)}{2} - 45 = 63 + 28 - 45 = 46.$$
$$U_2 = (7)(9) - 46 = 17.$$

3. Choose the lesser value of U, either U_1 or U_2 for significance testing; in this case we choose U_2. The critical value for df = 7, 9 is 12. Since U_2 is not less than the critical value at $\alpha = 0.05$ we do not reject the null hypothesis, and conclude that the children who stay at home do not differ in their age.

7.3 The Kruskal–Wallis for a one-way, model I ANOVA design

This test rests on the same principles as does the Mann–Whitney U test, except that it applies to more than two groups: if the null hypothesis that the groups were obtained from the same population is true, then the observations from all groups would be expected to be ranked randomly, as opposed to the observations from each group being ranked together. The assumption of independence among the groups holds, as does the assumption that ties are due to measurement error. Indeed, if there are ties, the computation of the statistic is slightly different, as will be seen below. The test can be applied to samples of equal or different size. Each group's sample size will be denoted as n_i, where i ranges from one to a, where a is the number of groups compared. The sum of all sample sizes will be referred to as $\sum\sum n$.The test proceeds as follows:

1. The observations of all samples are ranked together. Ranking is done taking into consideration the sign of the observation.
2. The sample size and the sum of the ranks $\sum R_1$ for all the groups are computed.
3. The formula for the H statistic for the Kruskal–Wallis test is shown in formula 7.2. This formula is to be used if there are no ties in the data. The H statistic is compared with the critical value at df = $a - 1$ of the χ^2 distribution. Selected critical values of the χ^2 distribution are shown in Table 7.2.

Formula 7.2

The H statistic for the Kruskal–Wallis test (no ties).

$$H = \left[\left(\frac{12}{(\sum\sum n)(\sum\sum n+1)}\right)\left(\sum\frac{(\sum R_i)^2}{n_i}\right)\right] - 3\left(\sum\sum n+1\right)$$

Compared with the χ^2 distribution with df = a − 1.

Table 7.2 Selected critical values of the chi-square distribution.

Degrees of freedom (df)	$p = 0.05$	0.02	0.01
1	3.841	5.412	6.635
2	5.991	7.824	9.210
3	7.815	9.837	11.341
4	9.488	11.668	13.277
5	11.070	13.388	15.086

If there are ties in the ranking procedure, than H must be divided by a correction term D. For example:

$$\left.\begin{matrix}1\\1\\1\end{matrix}\right\} t_1 = 3$$

$$\left.\begin{matrix}2\\2\end{matrix}\right\} t_2 = 2$$

$$\begin{matrix}3\\4\end{matrix}$$

$$\left.\begin{matrix}5\\5\\5\\5\end{matrix}\right\} t_3 = 4$$

In this sample, there are 11 observations and there are 3 groups of tied variates. In the first group, there were 3 tied variates. Therefore, $t_1 = 3$. In the second and third groups there were 2 and 4 tied variates, so $t_2 = 2$ and $t_3 = 4$ respectively. The correction for the H statistic considers the number of tied variates, and the number of groups of tied variates, by computing for each group the function T_j. The complete formula for the correction term is shown in formula 7.3.

Formula 7.3

The H statistic for the Kruskal–Wallis test when ties occur.

$$H = \frac{H}{D}, \text{ where}$$

$$D = 1 - \frac{\sum T_j}{\left(\sum\sum n_i - 1\right)\left(\sum\sum n_i\right)\left(\sum\sum n_i + 1\right)}$$

where

T is computed for each group as $(t_j - 1)(t_j)(t_j + 1)$, where

$\sum T_j$ is the sum of T across all groups of tied variates, and

$\sum\sum n_.$ is the sum of all sample sizes.

We now practice the computation of T and $\sum T_j$ for:

$$\left.\begin{matrix}1\\1\\1\end{matrix}\right\} t_1 = 3 \quad T_1 = (t_j - 1)(t_j)(t_j + 1) = (3 - 1)(3)(3 + 1) = 24$$

$$\left.\begin{matrix}2\\2\end{matrix}\right\} t_2 = 2 \quad T_2 = (t_j - 1)(t_j)(t_j + 1) = (2 - 1)(2)(2 + 1) = 6$$

3

4

$$\left.\begin{matrix}5\\5\\5\\5\end{matrix}\right\} t_3 = 4 \quad T_3 = (t_j - 1)(t_j)(t_j + 1) = (4 - 1)(4)(4 + 1) = 60.$$

For this small data set, then

$$D = 1 - \frac{\sum T_j}{\left(\sum\sum n_i - 1\right)\left(\sum\sum n_i\right)\left(\sum\sum n_i + 1\right)}$$

$$D = 1 - \frac{24 + 6 + 60}{(11 - 1)(11)(11 + 1)} = \frac{90}{1320} = 1 - 0.068 = 0.932.$$

We now practice the computation of the Kruskal–Wallis test with the same study which tested the hypothesis that children from two different ethnic groups are of the same age. However, we now expand our study to include a third ethnic group. The data set is called "children's economic contribution three groups." The null hypothesis is that the three groups do not differ for age.

We follow the already mentioned steps:

1. The observations of all samples are ranked together.

Age	Group	Rank
4	1	17.0
3	1	5.0
4	1	17.0
3	1	5.0
4	1	17.0
4	1	17.0
4	1	17.0

(cont.)

Age	Group	Rank
5	2	26.0
5	2	26.0
4	2	17.0
6	2	28.5
3	2	5.0
4	2	17.0
3	2	5.0
5	2	26.0
6	2	28.5
3	3	5.0
4	3	17.0
3	3	5.0
4	3	17.0
3	3	5.0
3	3	5.0
3	3	5.0
4	3	17.0
4	3	17.0
4	3	17.0
4	3	17.0
4	3	17.0
4	3	17.0

2. The sample size and the sum of the ranks $\sum R_i$ are computed for all the groups.

Group 1	Group 2	Group 3	
$n_1 = 7$	$n_2 = 9$	$n_3 = 13$	Therefore: $\sum\sum n = 29$
$\sum R_1 = 95$	$\sum R_2 = 179$	$\sum R_1 = 161$	

3. The H statistic is computed as follows:

$$H = \left[\left(\frac{12}{(29)(30)}\right)\left(\frac{95^2}{7} + \frac{179^2}{9} + \frac{161^2}{13}\right)\right] - 3(30)$$
$$= [(0.013793)(6843.3199)] - 90 = 4.39.$$

This H statistic must be corrected because the data set has ties. There are four groups of tied variates, for each of which we compute T.

Age	Group	Rank	
3	1	5.0	
3	1	5.0	
3	2	5.0	
3	2	5.0	
3	3	5.0	$T_1 = 9 = 8{*}9{*}10 = 720$
3	3	5.0	
3	3	5.0	
3	3	5.0	
3	3	5.0	
4	1	17.0	
4	1	17.0	
4	1	17.0	
4	1	17.0	
4	1	17.0	
4	2	17.0	
4	2	17.0	
4	3	17.0	$T_2 = 15 = 14{*}15{*}16 = 3{,}360$
4	3	17.0	
4	3	17.0	
4	3	17.0	
4	3	17.0	
4	3	17.0	
4	3	17.0	
4	3	17.0	
5	2	26.0	
5	2	26.0	$T_3 = 3 = 2{*}3{*}4 = 24$
5	2	26.0	
6	2	28.5	
6	2	28.5	$T_4 = 2 = 1{*}2{*}3 = 6$

$$D = 1 - \frac{720 + 3360 + 24 + 6}{28{*}29{*}30} = 1 - \frac{4110}{24{,}360} = 1 - 0.1687 = 0.8313.$$

$$H = \frac{H}{D} = \frac{4.39}{0.8313} = 5.281.$$

The degrees of freedom are 2, and the critical value at 0.05 is 5.991. Therefore, we accept the null hypothesis. The groups do not differ in the age at which children are expected to make economic contributions to the households in terms of gathered foods.

If a Kruskal–Wallis test is performed and the null hypothesis is rejected, the investigator needs to determine which group(s) is different from which. Generally, this test will simply be the Mann–Whitney U test which we just learned (see Sokal and Rohlf 1995).

Practice problem 7.3

We now practice the Kruskal–Wallis test with an archaeological data set collected at the Van Horn Creek site consisting of the weight in grams of the bottom valve of oysters (White, work in progress). The data shown here come from three stratigraphic levels. The raw data are in the data set called "Rangia shells in three levels." The data shown here include the raw data and the final rank, that is, the rank after ties were assigned. The null hypothesis is that the shells from the three levels do not differ in their weight.

The usual steps are followed:

1. Rank the observations of all samples together.

Obs.	Weight	Level	Rank
1	6.5	3	1.0
2	8.4	2	2.0
3	9.1	1	3.0
4	17.1	3	4.0
5	18.6	3	5.0
6	19.3	3	6.0
7	21.2	1	7.5
8	21.2	3	7.5
9	22.0	3	9.0
10	25.6	2	10.0
11	26.1	1	11.0
12	30.0	1	12.0
13	35.2	1	13.0
14	36.4	3	14.0
15	39.7	3	15.0
16	40.9	1	16.0
17	41.8	2	17.0
18	45.1	3	18.0
19	55.0	1	19.0
20	55.8	2	20.0
21	58.9	2	21.0
22	61.5	1	22.0
23	67.5	2	23.0
24	68.9	1	24.0
25	74.7	1	25.0
26	78.1	3	26.0
27	89.3	3	27.0
28	93.0	2	28.0
29	112.1	1	29.0

2. Compute the sample size and the sum of the ranks $\sum R_i$ for all groups.

Group 1	Group 2	Group 3	
$n_1 = 11$	$n_2 = 7$	$n_3 = 11$	Therefore: $\sum\sum n = 29$
$\sum R_1 = 181.5$	$\sum R_2 = 121$	$\sum R_3 = 132.5$	

3. Compute the H statistic:

$$H = \left[\left(\frac{12}{(29)(29+1)}\right)\left(\frac{(181.5)^2}{11} + \frac{(121)^2}{7} + \frac{(132.5)^2}{11}\right)\right] - 3(29+1) =$$

$$H = (0.013793)(6{,}682.344) - 90 = 2.17.$$

Because there was one tie, the H statistic must be corrected by dividing it by D. We first compute T_j for the only group of tied variates, that is, for:

$$\left.\begin{array}{l} 21.2 \\ 21.2 \end{array}\right\} t_1 = 2. \ T_1 = (2-1)(2)(2+1) = 6.$$

$$D = 1 - \frac{6}{(29-1)(29)(29+1)} = \frac{6}{24{,}360} = 1 - 0.000246 = 0.99975.$$

Last, we compute the corrected H, and compare it with the χ^2 table with degrees of freedom $= a - 1$, or $3 - 1 = 2$. $H = \frac{H}{D} = \frac{2.17}{0.99975} = 2.151$. Since this H statistic is less than the critical value at $\alpha = 0.05$ (CV $= 5.991$), we fail to reject the null hypothesis. The samples of oysters appear to have been obtained from a single population.

7.4 The Wilcoxon signed-ranks test for a two-sample paired design

The last non-parametric test covered in this chapter is one to be used instead of a paired t test. The reader recalls that an assumption of a paired t test is that the differences be normally distributed. If this assumption is violated, or if the data set is very small, or the measures are approximate, then a non-parametric test should be used. As in the paired t test, the two-tailed null hypothesis is that the differences as a whole are not significantly different from 0. However, the null hypothesis is not expressed in parametric terms. Frequently the H0 will be stated in a one-tailed manner, since the researcher has reason to know that the treatment will only affect the subjects in one direction. The test does assume that each individual was randomly selected, and that ties are the result of imperfect measures.

The Wilcoxon signed-ranks test computes differences between the two measures taken on the individual (or between the measures taken on matched individuals, whatever the case may be). Such differences are then ranked *without regard to the sign* (*although the sign of the difference will be used later*). That is, −1 and 1 would be tied and thus ranked equally.

Two extreme cases may occur: if the treatment causes all subjects to change either positively or negatively, then all the differences will go in one direction. If, however, the

Table 7.3 Selected critical values of T for the Wilcoxon signed-ranks test.

n	0.05	0.02	0.01
8	3	1	0
10	8	5	3
11	10	7	5
15	25	19	15
20	52	43	37
25	89	76	68
30	137	120	109

treatment does not affect the subjects, some differences will go in one direction, some in the other, and some will be equal to 0. The test proceeds as follows:

1. The difference between the two measures is taken.
2. The differences are ranked without regard to the sign, although the sign will be needed later.
3. The ranks of the negative ($\sum R-$) and the positive ($\sum R+$) differences are added up separately. If some of the differences are 0 they should be equally divided between the positive and the negative differences. In case there is an odd number of 0 differences, one of the differences should be discarded, *thus reducing the sample size by one*, and the others should be equally divided. You should be aware that computer packages differ on how they handle the issue of differences whose value is 0. Therefore, you may obtain output from different packages which print different sample sizes because some dropped while others did not drop 0 differences.
4. The smaller sum of the ranks is called T, and is compared with the critical values provided by Table 7.3. **If the sample T is less than or equal to the table's value, then the null hypothesis is rejected**. The table functions in this manner because if the treatment affected all subjects in one direction, then all the differences will be either positive or negative. Therefore, the smaller sum of ranks will be 0, since all of the differences will have the other sign. Hence, the smaller the value of T, the more different the two groups will be.

We will practice this test with the following example: an anthropologist is interested in the stability of prestige in a community. He interviews a member of the community involved in religious activities, and who is outside the community's prestige hierarchy, and gets the informant's view of the hierarchy. The researcher asks the subject to rank order from the highest to the lowest prestige eight heads of household. The anthropologist returns a year later, and asks the same informant to rank the same individuals. The null hypothesis is that there has been no change. Therefore, this is a two-tailed test. The data are below:

Individual ranked	Rank first year	Rank second year	Difference (first − second)
1	2	3	−1
2	7	7	0
3	8	8	0
4	1	2	−1
5	4	6	−2
6	5	4	1
7	6	5	1
8	3	1	2

We rank the differences from lowest to highest, and assign ties to them when appropriate.

Difference	Initial rank	Final rank
0	1	1.5
0	2	1.5
1	3	4.5
1	4	4.5
−1	5	4.5
−1	6	4.5
−2	7	7.5
2	8	7.5

1. We add up the ranks of positive and negative differences separately. The ranks of both zero differences are equally divided between the positive and negative sum of ranks.

$$\sum R+ = 1.5 + 4.5 + 4.5 + 7.5 = 18, \ \sum R- = 1.5 + 4.5 + 4.5 + 7.5 = 18.$$

2. We choose the smaller sum (in this case either sum), and compare it to the critical value for $n = 8$ at $\alpha = 0.05$ (CV = 3). We do not reject the null hypothesis, since our T is greater than the critical value.

Practice problem 7.4

An anthropologist is interested in determining if in a particular community, the menarcheal age has changed from one generation to the other. She decides to compare the menarcheal age of mothers and daughters, which makes this a matched design. In this community, however, the Western-style calendar is not used, making it difficult to ascertain the exact menarcheal age. The anthropologist estimates as best as possible the menarcheal age, in consultation with the subjects and other related females. Because the data are approximations, and because of the small sample size, a non-parametric test should be used. Test the hypothesis that the menarcheal age has not changed.

Mothers' menarcheal age	Daughters' menarcheal age	Difference
15	13	2
14	14	0
12	11	1
14	13	1
11	11	0
16	14	2
15	13	2
16	15	1
14	12	2
11	13	−2
12	12	0
13	12	1

We rank the differences from lowest to highest and assign ties to them when appropriate.

Difference	Initial rank	Final rank
0	1	2
0	2	2
0	3	2
1	4	5.5
1	5	5.5
1	6	5.5
1	7	5.5
−2	8	10
2	9	10
2	10	10
2	11	10
2	12	10

1. We add up the ranks of positive and of negative differences separately. Since there are three zero differences, we drop one (thus $n = 11$ now), and divide the other two equally into both groups:

$$\sum R+ = 2 + 5.5 + 5.5 + 5.5 + 5.5 + 10 + 10 + 10 + 10 = 64.$$

$$\sum R- = 2 + 10 = 12.$$

2. We choose the smaller sum, namely, 12, and compare it with the critical value for $n = 11$. The critical value at $\alpha = 0.05$ is 10, so we accept H0. We conclude that mothers and daughters did not experience a significantly different menarcheal age.

Anthropology and non-parametric tests

Even I, with my limited exposure to television, know that forensic science is a favorite topic in popular TV shows. Something that is featured prominently in these shows is the facial reconstruction of skeletonized human remains using either clay or computer-generated images. An important question is how (based on that type of data) those computer-generated images should be generated. If there are more than one approach available for such reconstruction, do they lead researchers to reconstruct a face with essentially the same features? De Greef *et al.* (2005) ask to what extent a mobile ultrasound (US) system and a CT-based technique compare with each other in terms of reconstructing human faces. Because the authors have small sample sizes ($n = 12$) they apply a Wilcoxon paired signed-rank test to measures on the same individual taken with both apparatuses. The authors show that only 5.8% of all 52 measures compared differ significantly at the 0.01 level (De Greef *et al.* 2005).

The relationship between geography, isolation, and genetic heterogeneity (measured by mean heterozygosity) is of interest to anthropologists who want to understand the evolution of native populations in South America. Kohlrausch *et al.* (2005) study the influence of geography on the level of heterozygosities in nine Amerindian groups. The authors note that since mean heterozygosities do not follow a normal distribution the data need to be analyzed with a non-parametric test. The null hypothesis that the heterozygosities did not differ among all groups was rejected, and followed up with a non-parametric two-sample test which tested each sample against the others. The authors report that the lowest heterozygosity levels are found in three groups which have been particularly isolated. This is expected because these groups would not be likely to receive much gene flow (Kohlrausch *et al.* 2005).

To what extent will humans go to embellish themselves? A simple look around you will tell you that we like to color or remove our hair, to permanently mark our skin with tattoos or scarification, to pierce almost anything, etc. What about intentional head deformation? This is not something we see too often in living people (nor am

I advocating that anthropology students do this to their babies!), but it is something practiced in many cultures. Anthropologists are interested in determining the type of cranial deformation practiced in specific cultures, given that some practiced only one, while others practiced several (Arnold *et al.* 2008). Working with a sample from Eastern Europe, Arnold *et al.* (2008) determined that only one type of cranial deformation (tabular fronto-occipital) was practiced in this region. In addition, the authors determined that while the deformation affected the neurocranium, it hardly affected the facial cranium. A non-parametric Mann–Whitney U test was used in this project because the authors had a small sample size at their disposal (only 11 artificially deformed skulls).

7.5 Chapter 7 key concepts

- Non-parametric tests vs. parametric tests.
- When should a non-parametric test be considered?
- Ranking with ties.
- Ranking considering the sign of the digit or with the absolute value.
- Review why the decision rule in the statistical tables used in this chapter differ from those used previously.
- Next time you see a research paper which applies a *t* test or ANOVA or a paired *t* test, ask yourself if the authors justified this choice, or if they should have used a non-parametric test.

7.6 Computer resources

1. I prefer the PASW output over the output produced by SAS for Mann–Whitney. PASW computes the U statistic and the sum of the ranks of the first group exactly as we did here. Both computer packages also transform the statistic into a *z* score and produce the probability associated with that *z* score so that you don't have to look at the table of critical values. SAS does not compute the U statistic, but only the *z* score (and a chi-square approximation).
2. I prefer the SAS output over the PASW output for Kruskal–Wallis. While the former prints the sum of the ranks of each group, the latter only gives you the test statistic and the *p* value. Both packages print a chi-square approximation so if you want to check that you computed your H statistic correctly you cannot do it with either program.

7.7 Chapter 7 exercises

All data sets are available for download at www.cambridge.org/9780521147088.

1. A social worker is interested in determining if the age of individuals in a nursing home is different between two ethnic groups. Because of the small size of the data

set "nursing home ages," use a Mann–Whitney U test to determine if the groups differ significantly in their ages.

2. A medical anthropologist is interested in comparing the scores of three ethnic groups on a depression screening scale. She suspects that one of the ethnic groups has fewer depressive symptoms and is scoring differently than the others. Use a Kruskal–Wallis test to determine if the groups differ in their depression screening scale. The data set is called "depression."
3. A community center has hired a social worker with the purpose of educating the community on better eating habits. One of the main concerns of the center is the high consumption of eggs in the traditional diet of the community. The social worker chooses to work with ten homemakers for a period of 4 weeks, and collects the self-reported number of eggs per weeks consumed in the household before and after the educational campaign. Test the null hypothesis that there has been no change in the self-reported number of eggs with a Wilcoxon-signed rank test. The data set is called "eggs."

8 The analysis of frequencies

This chapter presents a departure from the statistical tests covered so far: we do not deal with quantitative variables such as fertility, prestige ranks, height, weight, etc., within a sample or among several samples. Instead we deal with frequencies of occurrences or events. We work with variables which have a few possible outcomes, and we are concerned with the number of individuals who have each of the outcomes (where *a* is the number of outcomes). Often-used anthropological frequency data are gene frequencies, the frequencies of males and females, the frequencies of different ethnic groups, the frequencies of different pottery styles in archaeological sites, etc.

8.1 The X^2 test for goodness-of-fit

In this section we discuss the well-known and widely used chi-square test (X^2), applied as a goodness-of-fit test and as a test for independence of two variables. Not only is the X^2 test widely used in anthropology, but unfortunately it has also been widely misused. The most common error about the X^2 test is its own name: it is frequently referred to as χ^2, which is the parametric notation for a theoretical frequency distribution against which the statistic computed from the sample (X^2) should be compared. Therefore, the test statistic is X^2, while the theoretical distribution is χ^2. In addition, a correction which must be applied to small data sets is also discussed.

The purpose of the goodness-of-fit test is to determine if the observed frequencies of events depart significantly from expected frequencies proposed by a null hypothesis. How that null hypothesis is generated depends on the specific research project. Usually, the expected frequencies (f_e) are generated in either of two ways: (1) The frequencies of observations in all outcomes are expected to be the same, that is, there is an equal probability expected in the various outcomes, or (2) the frequencies are generated by an expectation based on prior knowledge of the data's nature. An example of the first case would be the comparison of the number of observed male and female students in (e.g.) medical school, where it is expected that half the students would be males, and half would be females. If we sample $n = 70$ students, then the expected frequency of male students would be $f_{males} = \frac{70}{2} = 35$, and the expected frequency of female students would be $f_{females} = \frac{70}{2} = 35$. An example of the second null hypothesis generation is when we compare observed gene frequencies with those expected under the assumptions of the Hardy–Weinberg equilibrium. That is, if the population is not evolving and its members mate randomly, then we would expect to see specific frequencies of phenotypes. We use a X^2 to compare such expected frequencies with the observed ones.

However the null hypothesis' expected frequencies are generated, the mathematical calculations for the computation of the X^2 test are the same. The test is not parametric, since it does not make statements about the parameters or distribution of the population from which the sample was obtained. All the *null hypothesis* states is that H0: $f_o = f_e$, that is, that the observed frequencies are not significantly different from those expected if the null hypothesis is true. Although non-parametric, the X^2 still assumes random sampling and that the expected frequencies are at least five. The first assumption is a matter of the research design. The second assumption must be attended to: if it is violated, the computation of X^2 must be corrected. This will be covered later in the chapter. The chi-square test proceeds as follows:

1. The null hypothesis is stated, and the expected frequencies are generated. Usually the researcher states how the expected frequencies were computed, if by an expectation of equality among all outcomes, or by an expectation based on previous studies or by a theoretical expectation (like the Hardy–Weinberg equilibrium).

2. The data are arranged so that the observed and expected frequencies are next to each other. For example, if we collected a sample of 70 medical school students, and want to test the null hypothesis that the frequency of males is the same as the frequency of females (that is, H0: $f_{females} = f_{males} = \frac{70}{2} = 35$), and observe 30 females and 40 males, then we would arrange the data as follows:

	females	males
observed (f_o)	30	40
expected (f_e)	35	35

3. We compute the X^2 statistic as shown in formula 8.1.

Formula 8.1

Formula for the computation of the X^2.

$$X^2 = \sum \frac{(f_o - f_e)^2}{f_e}, \quad \text{across all } a \text{ outcomes, with}$$

$$\text{df} = a - 1.$$

4. If our X^2 is equal to or greater than the critical value provided by the chi-square table with df $= a - 1$, then we reject the null hypothesis. Table 8.1 shows selected values of the χ^2 distribution (again, please keep in mind that the statistic we compute is not a chi, but that the distribution with which we compare it is).

We first practice with the data presented above on the gender of a sample of 70 medical students. The statistic is $X^2 = \frac{30-35^2}{35} + \frac{40-35^2}{35} = 0.7143 + 0.7143 = 1.4286$. Since there are two categories (males and females) then df $= 2 - 1 = 1$. Because the

Table 8.1 Selected critical values of the chi-square distribution.

Degrees of freedom (df)	$p =$ 0.05	0.02	0.01
1	3.841	5.412	6.635
2	5.991	7.824	9.210
3	7.815	9.837	11.341
4	9.488	11.668	13.277
5	11.070	13.388	15.086

critical value at 0.05 is 3.841, we accept the null hypothesis that the observed frequencies of male and female students do not differ from those expected under the null hypothesis.

It is easy to grasp how the chi-square statistic captures the difference between expected and observed frequencies. If instead of observing 30 females and 40 males we had observed 60 females and 10 males the statistic would have been:

$$X^2 = \frac{(60 - 35)^2}{35} + \frac{(10 - 35)^2}{35} = 17.85 + 17.85 = 35.71.$$

In this case we would have rejected the null hypothesis with a great degree of confidence. Therefore, the larger the difference between the expected and observed frequencies, the more likely we are to reject the null hypothesis.

Practice problem 8.1

We now practice the goodness-of-fit X^2 test with an example in which the null hypothesis is generated according to a previous study. Earlier in this book, we computed the frequencies of ceramic types from test unit A in Depot Creek shell mound (data from White 1994; Table 3.1). We can use these frequencies to generate the null hypothesis for another test unit. That is, we can test the hypothesis that there is no difference between the observed (test unit B) and the expected frequencies (generated from test unit A). The data from test unit A are:

Ceramic types from Depot Creek shell mound (86–56). Test unit A.

Unit A	Frequency	Percent	Cumulative frequency	Cumulative percent
Check-stamped	185	58.18	185	58.18
Grog-tempered	9	2.83	194	61.01
Indent-stamped	28	8.80	222	69.81
Other (Comp & Cord)	9	2.83	231	72.64
Sand-tempered	44	13.84	275	86.48
Simple-stamped	43	13.52	318	100.00

Note that two categories (comp-stamped and cord-stamped) have been merged into one ("other") because their individual contribution to the total sample size was small. If they had been left separately, the assumption that the minimum expected frequency per category be 5 would have been violated. With these data from test unit A, we can generate expected frequencies for test unit B. Since 58.18% of the ceramics in test unit A are check-tempered, we expect 58.18% of the ceramics in test unit B to be check-stamped. To obtain the expected frequencies we multiply the sample size of test unit B ($n = 216$) by the percentages from test unit A. For example, for check-stamped: $f_e = (216)(0.5818) = 125.67$. We compute the other expected frequencies in the same manner. We denote observed frequencies as f_o and expected frequencies as f_e.

Ceramic types from Depot Creek shell mound (86–56). Test unit B.

Ceramic type	f_o	f_e	$f_o - f_e$	$\frac{f_o - f_e^2}{f_e}$
Check-stamped	88	(216)(0.5818) = 125.67	−37.67	11.29
Grog-tempered	15	(216)(0.0283) = 6.11	8.89	12.93
Indent-stamped	46	(216)(0.088) = 19.01	26.97	38.32
Other	12	(216)(0.0283) = 6.11	5.89	5.68
Sand-tempered	40	(216)(0.1384) = 29.90	10.10	3.41
Simple-stamped	15	(216)(0.1352) = 29.20	−14.20	6.90
	$n = 216$		$\sum \approx 0$	$\sum = 78.53.$

Since there are 6 categories, then df = 5. We reject the null hypothesis, because our X^2 is greater than the critical value at 0.05 (11.07) and even at 0.001 (15.086). We conclude that the two test units differ in their frequencies of ceramic types.

Before we continue, the reasoning behind the computation of the X^2 formula should be discussed, specifically, why we square the deviation between the observed and expected frequencies and divide it by the expected frequencies. Notice that the X^2 statistic computes the "raw" difference between the observed and expected frequencies. The difference then must be squared so that the positive and negative differences do not cancel each other out (indeed, if the "raw" differences from the previous example are added, they sum to 0, as shown above). The reason these squared deviations ought to be divided by the expected frequency is that by dividing the difference, the difference is expressed as a proportion of the expected frequency. Therefore, if the squared deviation is virtually equal to the expected frequency, then we would be dividing a number basically by itself, so the quotient of the division would be close to 1. In this latter situation, we are not likely to reject the null hypothesis. Now, if the squared deviation is a large number because there is a large difference between f_o and f_e, then we would be dividing a large number by a smaller one, the quotient of this division would be much greater than one. In this latter situation, we would be likely to reject the null hypothesis. Finally, please note that since we work with squared deviations, the χ^2 distribution is not symmetrical, but cuts off at 0. Figure 8.1 shows the χ^2 distribution for several degrees of freedom.

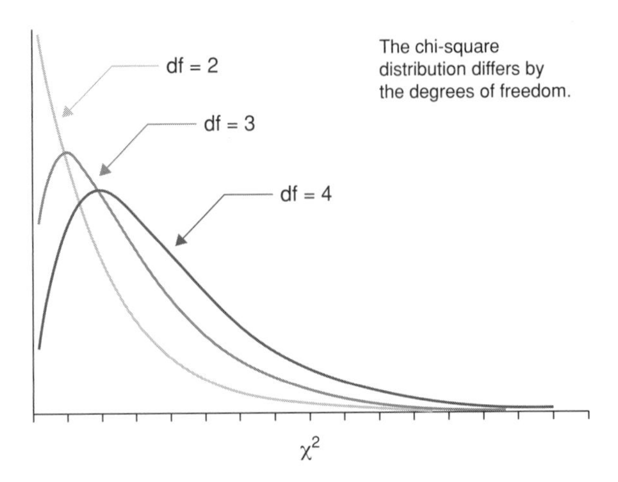

Figure 8.1 The chi-square distribution.

8.2 The Kolmogorov–Smirnov one sample test

The Kolmogorov–Smirnov (K–S) one sample test is quite useful for testing that a set of observed frequencies does not differ from expected ones when the observed frequencies are arranged in ordered cells. Up to now we have worked with ceramic types which have no obvious order. But what about if we collect data on agricultural yield in plots of land that are closer and closer to the water source and thus receive better irrigation? Or what about if we count the number of deaths per month in a region which experiences climatic changes during the year? The frequency of flu deaths is higher during the winter months than during the warmer months, so there is an order to the observed number of deaths throughout the year. The Kolmogorov–Smirnov test is the right test to apply when you collect frequencies observed in ordered categories. Fortunately the expected frequencies are computed as if we were doing a regular X^2. Moreover, the test is quite easy to compute by hand, although tables of its critical values are slightly more complicated than the usual statistical table because the sample size n has to be a multiple of the number of categories (a). If it is not, the analyst may use the critical values for the more conservative, lower, appropriate n. Of course, with computers this is not a concern. Table 8.2 reproduces a few selected values of the complete K–S table.

To compute the K–S test we take the difference between the observed and the expected frequencies and choose the largest *absolute* difference. This is what we call d_{max} and it is the test statistic, which is compared with the critical value in the table at 0.05 for the appropriate sample size and number of categories. If our value is equal to or greater than the critical value, we reject the null hypothesis.

Let us consider a research project on the number of births in a small village clinic, in which births are recorded in a year divided in quarters. The purpose of the project is to determine if the number of births is distributed equally across the four quarters. We recorded 10 births in the first quarter, 12 in the second, and so forth. The expected frequencies were generated by dividing the total sample size ($n = 47$) by four. The null

Table 8.2 Selected critical values for d_{max} for the Kolmogorov–Smirnov goodness-of-fit test for discrete data.

a	n	0.05	0.02
4	36	7	8
4	40	8	9
4	44	8	9
4	48	8	10
4	64	9	11
4	68	10	11

hypothesis is that the observed frequencies of births do not differ from those expected by an equal distribution across the four quarters. The critical value for $a = 4$ and $n = 47$ is not in the table (47 is not a multiple of four), so we use the more conservative value at df = 4, 44, which is CV = 8. Our d_{max} is only 3.25, and not significant. We do not reject the null hypothesis that births occur at an equal frequency in the four quarters.

Quarter	f_o	$f_e\,(\frac{47}{4})$	$f_o - f_e$	$\vert f_o - f_e\vert$
1	10	11.75	−1.75	1.75
2	12	11.75	0.25	0.25
3	15	11.75	3.25	**3.25**
4	10	11.75	−1.75	1.75

Practice problem 8.2

Let us consider the following data, collected at a similar clinic but in an environment with strong food and water supply seasonal variation. Is the frequency of births the same across the four quarters?

Quarter	f_o	$f_e(\frac{67}{4})$	$f_o - f_e$	$\vert f_o - f_e\vert$
1	10	16.75	−6.75	6.75
2	12	16.75	−4.75	4.25
3	15	16.75	−1.75	1.75
4	30	16.75	13.25	**13.25**

Our sample size is now 67 and we will need to use the approximate df = 4, 64, for which the critical value at $\alpha = 0.05$ is 9 and the critical value at $\alpha = 0.02$ is 11. Since our d_{max} is greater than 11, we reject the null hypothesis at the 0.02 level.

Table 8.3 The cells in a 2×2 table.

		Variable B	
Variable A	B1 (Column 1)	B2 (column 2)	Totals
A1 (row 1)	a	b	$\sum row\ 1 = a + b$
A2 (row 2)	c	d	$\sum row\ 2 = c + d$
Totals	$\sum column\ 1 = a + c$	$\sum column\ 2 = b + d$	Grand total $= a + b + c + d = n$
In **epidemiological terminology** this table may be better understood as follows:			
		Exposure to risk	
Disease presence	Exposed	Unexposed	Totals
Positive	a	b	$\sum row\ 1 = a + b$
Negative	c	d	$\sum row\ 2 = c + d$
Totals	$\sum column\ 1 = a + c$	$\sum column\ 2 = b + d$	Grand total $= a + b + c + d = n$
In **medical terminology** this table may be better understood as follows:			
		Outcome	
Treatment	Success	Failure	Totals
Treatment 1	a	b	$\sum row\ 1 = a + b$
Treatment 2	c	d	$\sum row\ 2 = c + d$
Totals	$\sum column\ 1 = a + c$	$\sum column\ 2 = b + d$	Grand total $= a + b + c + d = n$

8.3 The X^2 test for independence of variables

The X^2 test for independence of variables does not differ from the goodness-of-fit test in its null hypothesis or even computation. The purpose is still to determine if the observed frequencies of observations falling into specific outcomes are significantly different from those proposed by the null hypothesis. Nonetheless, the outcomes are now determined by two variables, each of which has at least two possible outcomes. Therefore, the *null hypothesis* may now be re-worded as H0: the two variables are independent. Thus, we could test if the number of male and female medical students is independent of socioeconomic status (defined as a discrete variable, with three categories: high, low, and middle). Or, we could test that the frequency of six ceramic types is independent from the stratigraphic level from which they were obtained. Or, we could test that the frequency of several hemoglobin types does not differ across two different ethnic groups, etc. The data are usually arranged in a table with r rows and c columns, where the number of individuals which belong to each cell (defined by the two categories) is displayed. Table 8.3 shows a table with the frequency of outcomes of two variables. This is called a 2×2 table because both variables have two outcomes. The table shows the outcomes described in frequently used terminology in the medical and public health fields.

The H0's expected frequencies are usually generated as if the two variables were independent of each other. This is done by multiplying a cell's column (column total or

Table 8.4 Computing the expected frequencies in a X^2 test for independence of variables.

	Variable B		
Variable A	B1	B2	Totals
A1	$f_e = \frac{a+b^{*}a+c}{n}$	$f_e = \frac{a+b^{*}b+d}{n}$	$\sum row\ 1 = a+b$
A2	$f_e = \frac{c+d^{*}a+c}{n}$	$f_e = \frac{c+b^{*}b+d}{n}$	$\sum row\ 2 = c+d$
	$\sum column\ 1 = a+c$	$\sum column\ 2 = b+d$	Grand total $= a+b+c+d=n$

CT) and row totals (or RT), and dividing by the total sample size or grand total. The computation of the expected frequencies is shown in Table 8.4.

The formula for the computation for the X^2 statistic remains the same: we square the difference between the observed and expected frequencies, divide by the expected value, and sum across all cells. The formula for the degrees of freedom takes into consideration the number of columns and rows so that df $= (c-1)(r-1)$.

Let us presume that an anthropologist is interested in determining if in a particular culture, female property ownership is related to the female's marital status. The researcher collects data consisting of the number of subjects who are married and unmarried, and who do and do not own property. The data are below:

First variable: marital status

		Married	Unmarried	Total
Second variable:	Own property	40	6	$40+6=46$
property ownership	Do not own property	20	10	$20+10=30$
		$40+20=60$	$6+10=16$	$n=60+16=76$
				or $n=46+30=76$

The first step is to compute the expected frequencies as follows:

First variable: marital status

		Married	Unmarried	
Property ownership	Own property	$\frac{46^{*}60}{76} = 36.32$	$\frac{46^{*}16}{76} = 9.68$	46
	Do not own property	$\frac{30^{*}60}{76} = 23.68$	$\frac{30^{*}16}{76} = 6.32$	30
		60	16	$n=76$

Note that since the expected frequencies are additive, only some of them needed to be computed in this manner. For example, for the first column, after we computed the f_e for cell a (36.32), we could have obtained the f_e for cell c as $60 - 36.32 = 23.68$. Having computed the expected frequencies for the first column, we can obtain the frequencies for the second column by subtracting from the row totals the expected frequencies of the first column. Thus, the expected frequency for unmarried property owners is: 46 – 36.32 = 9.68. The expected frequency of unmarried non-property owners is 30 – 23.68 = 6.32. The H0 is $f_e = f_o$. We now compute our X^2 statistic:

$$X^2 = \frac{(40-36.32)^2}{36.32} + \frac{(6-9.68)^2}{9.68} + \frac{(20-23.68)^2}{23.68} + \frac{(10-6.32)^2}{6.32} = 4.49.$$

The degrees of freedom are (2 – 1)*(2 – 1) = 1. Since the critical value at $\alpha = 0.05$ is 3.841, we reject the null hypothesis that adult female property ownership is independent of marital status in this community: there is a significant association between both variables.

Practice problem 8.3

A medical anthropologist is interested in investigating if in the community she is studying, mothers who use traditional medicine to heal their children have significantly more or less formal education than mothers who do not. The anthropologist divides the level of education into three levels: low, medium, and high. Test the null hypothesis that use of traditional medicine is independent of educational level. The data are found in the Excel file called "Education."

The first step is to arrange the data into a table that includes the observed and the expected frequencies. The expected frequencies are written in bold face. Notice that they add up to the column and row totals. The H0 is $f_e = f_o$.

	Education level			
Use of traditional medicine	High	Low	Medium	Total
No	14	8	4	26
Expected frequencies	**9.21**	**11.92**	**4.87**	
Yes	3	14	5	22
Expected frequencies	**7.79**	**10.08**	**4.13**	
Total	17	22	9	48

$$X^2 = \frac{(14-9.21)^2}{9.21} + \frac{(8-11.92)^2}{11.92} + \frac{(4-4.87)^2}{4.87} + \frac{(3-7.79)^2}{7.79} + \frac{(14-10.08)^2}{10.08} + \frac{(5-4.13)^2}{4.13} = 8.59, \ df = (3-1)(2-1) = 2.$$

The critical values at $\alpha = 0.02$ and $\alpha = 0.01$ are 7.824 and 9.210 respectively. Thus, we reject the null hypothesis at an alpha level less than 0.02 but greater than 0.01. Our data indicate that use of traditional medicine is not independent of level of education.

8.4 Yates' correction for continuity

It was mentioned previously that the X^2 test assumes that the cell's expected frequencies be at least five. In cases in which this assumption does not hold, the X^2 statistic should be corrected. However, there is some debate about which correction is best, as well as about how closely the assumption itself should be followed. If no computer is available, the correction of choice for a 2×2 table with expected frequencies under five is Yates' correction for continuity. The main advantage of Yates' correction is its easy calculation, while its main disadvantage is that it is too conservative (Manly 1992). Fleiss *et al.* note that the correction has been found to be overly conservative (it reduces the power of the test), but still recommend its use (Fleiss *et al.* 2003). If a computer is available, a Fisher's exact test is preferred over a X^2 with a Yates' correction. The Fisher's exact test is discussed below.

The formula for Yates' correction involves a minor change from the usual X^2 computation: after the difference between the f_o and the f_e is obtained, 0.5 is extracted from the difference's **absolute** value. That is, if the difference were −2, then we would correct it as $2 - 0.5 = 1.5$. The reader can see that Yates' correction diminishes the difference between the observed and expected frequencies, making the test more conservative. To distinguish a corrected from a non-corrected X^2, the former is usually denoted as X_c^2. The formula for Yates' correction is shown in formula 8.2.

Formula 8.2

Yates' correction for continuity.

$$X_c^2 = \sum \frac{(|f_o - f_e| - 0.5)^2}{f_e}$$

with summation across the a categories, and df as before.

Let us practice the correction with the data set concerning the mothers' education and their use of traditional medicine, since there were two cells whose expected frequency was under 5: 4.87 and 4.13. The H0 is $f_e = f_o$. The X^2 is computed as follows:

$$X_c^2 = \frac{(|14 - 9.21| - 0.5)^2}{9.21} + \frac{(|8 - 11.92| - 0.5)^2}{11.92} + \frac{(|4 - 4.87| - 0.5)^2}{4.87}$$
$$+ \frac{(|3 - 7.79| - 0.5)^2}{7.79} + \frac{(|14 - 10.08| - 0.5)^2}{10.08} + \frac{(|5 - 4.13| - 0.5)^2}{4.13} = 6.56.$$

$$\text{df} = (3 - 1)(2 - 1) = 2.$$

We reject the null hypothesis at the α level of 0.05, but not at the level of 0.02. Our X^2 has diminished by almost two units. Clearly, Yates' correction is more likely to lead to the retention of the null hypothesis.

8.5 The likelihood ratio test (the G test)

I dare say that no other statistical test is used as much in anthropology as is the chi-square test. Most undergraduate anthropology students learned to use it when they took their biological anthropology laboratory with the purpose of testing if gene frequencies were at equilibrium. Perhaps the X^2 test has been so prevalent because of its ease of computation and its obvious meaning: the more different the observed and expected frequencies, the larger the probability that we will reject the null hypothesis. However, more and more authors are recommending that the G test should be favored over the chi-square for goodness-of-fit tests in general, whether we are considering one or two variables. While Sokal and Rohlf have been strong proponents of the G test (Sokal and Rohlf 1995), Zar (1999) notes that a preference for the G test is not universal. There is also some disagreement on how the test should be called, as it is referred to as the G, the G^2, or the maximum likelihood test (Zar 1999). Fortunately the computation of the expected frequencies is performed in the same manner as was done for the chi-square test.

Like the chi-square test, the G test compares observed with expected frequencies. However, the difference between observed and expected frequencies is evaluated using a maximum likelihood approach as opposed to a least-squares approach. How the differences are evaluated makes sense as much as it does for a chi-square test. If two frequencies are very similar (say 30 and 35) and we divide one by the other, then we will obtain a number not too different from one ($\frac{f_o}{f_e} = \frac{30}{35} = 0.85$). Instead, if the frequencies are quite different (say 60 and 35) and we divide one by the other, then we will obtain a number largely different from one ($\frac{f_o}{f_e} = \frac{60}{35} = 1.75$). Therefore, just as in the chi-square, the larger the differences between the observed and expected frequencies, the larger the test statistic. For the computation of the G test you will need to use the natural log key in your calculator (ln). The formula for the computation of the G test is shown in formula 8.3.

Formula 8.3

The G test.

$$G = 2\sum^{a} f_o{}^* ln \frac{f_o}{f_e}$$

With df = $a - 1$ where a is the number of categories. The G statistic is compared with the table of chi-square critical values.

Let us practice the computation of the G statistic with the data on ceramic types from two test units from the Depot Creek shell mound. The H0 is $f_e = f_o$.

Table 8.5 Observed ceramic type frequencies from Depot Creek shell mound (86–56) test unit B compared with expected frequencies generated following test unit A expectations.

Ceramic type	f_o	f_e	$\frac{f_o}{f_e}$	$ln\frac{f_o}{f_e}$	$f_o{}^*ln\frac{f_o}{f_e}$
Check-stamped	88	125.67	0.700	−0.3563	−31.3136
Grog-tempered	15	6.11	2.45499	0.8981	13.47
Indent-stamped	46	19.01	2.4198	0.8836	40.65
Other	12	6.11	1.96399	0.67498	8.10
Sand-tempered	40	29.90	1.33779	0.2910	11.64
Simple-stamped	15	29.20	0.513699	−0.66612	−9.99
	$n = 216$				$\sum f_o{}^*ln\frac{f_o}{f_e} = 32.51$

The G statistic is 2*32.51 = 65.03. Since there are six categories, then df = 5. We reject the null hypothesis, because our G is greater than the critical value at 0.05 (11.07) and even at 0.001 (15.086). We conclude that the two test units differ in their frequencies of ceramic types. You can see that we reached the same conclusion we did with the chi-square test.

Practice problem 8.4

Let us practice the computation of the G statistic with the data set on use of traditional medicine in a group of women of three different levels of formal education. The H0 is $f_e = f_o$. The data are:

	Education level			
Use of traditional medicine	High	Low	Medium	Total
No	14	8	4	26
Expected frequencies	**9.21**	**11.92**	**4.87**	
Yes	3	14	5	22
Expected frequencies	**7.79**	**10.08**	**4.13**	
Total	17	22	9	48
	35.42	45.83	18.9	100

$$G = [2]\left[\left(14^*ln\frac{14}{9.21}\right) + \left(8^*ln\frac{8}{11.92}\right) + \left(4^*ln\frac{4}{4.87}\right) + \left(3^*ln\frac{3}{7.79}\right) + \left(14^*ln\frac{14}{10.08}\right) + \left(5^*ln\frac{5}{4.13}\right)\right] = [2][4.5775] = 9.155.$$

The degrees of freedom are $c - 1^*r - 1 = 2^*1 = 2$. The critical values at $\alpha = 0.02$ and $\alpha = 0.01$ are 7.824 and 9.210 respectively. Thus, we reject the null hypothesis at an alpha level less than 0.02 but greater than 0.01. Our conclusions with the G and the chi-square tests are in agreement.

The assumption of a minimum value of five for the expected frequencies also holds for the G test. It has been proposed that the probability of committing a type I error when there are two categories is high in a G test. For that reason the Williams' correction (instead of Yates') is advocated to make the test more conservative when there are two categories or when the expected frequencies are under five (Fleiss *et al.* 2003).

Formula 8.4

The Williams' correction for the G test.

$$G_{adj} = \frac{G}{1 + \frac{a+1}{6^*n}}$$

with $df = a - 1$ where a is the number of categories. The G_{adj} statistic is compared with the table of chi-square critical values.

Let us compute G_{adj} for the G test which tested the null hypothesis that ceramic type is the same in both test units of the Depot Creek site (where $n = 216$ and $a = 6$). $G_{adj} = \frac{65.03}{1+\frac{6+1}{6^*216}} = \frac{65.03}{1+0.005401} = 64.681$. The test statistic is lower (more conservative). Still, we reach the same conclusion and reject the null hypothesis.

8.6 Fisher's exact test

Allow me to introduce this computationally complex test with a rather familiar concept to us: let us remember the meaning of the areas of the normal curve represented by columns B and C in Table 3.5. You recall that column B is the area of the normal curve between the mean and any z score, while column C is the area of the normal curve beyond the z score. Another way in which we could look at column C is that this is the area of the normal curve associated with values as extreme and more extreme as the z score. We could call the values of the z scores beyond "our" z score "worse" or "more extreme" or "less likely."

Let us remember that since the value of the C column associated with a z score of 1.96 is 0.025, we can tell that the probability associated with finding values as extreme and more extreme than 1.96 and -1.96 is $2^*0.025 = 5\%$ of the distribution. Fisher's exact test (I don't need to tell you who proposed it) asks what the probability is of getting a 2×2 table of observed frequencies (with specified row and column totals) AND the probability of getting "more extreme" tables *with the same row and column totals*. More extreme here means decreasing one of the diagonals of a 2×2 table until at least one of the elements is 0. For each of the tables between the observed table and the table with a 0 element, we compute an exact probability of obtaining it. To answer the question what is the probability of obtaining the observed table and all "more extreme" tables, we add the probabilities of all the tables. The final probability considers both tails (just as we considered the probabilities of -1.96 and 1.96).

The Fisher's exact test is *the test of choice* in a 2×2 table with small sample sizes. You do not have to worry about performing any corrections with it. However, the test is rather involved in its computation and without the help of calculators (or computers, of course!) it is difficult to perform by hand. The Fisher's exact test requires the computation of the factorial of the table cells, the row and column totals, and the sample size. Just to remind you, the factorial of 2 is $2! = 2^*1 = 2$, the factorial of 3 is $3! = 3^*2^*1 = 6$, the factorial of $4! = 4^*3^*2^*1 = 24$, and so forth. Fortunately most hand-held calculators have a factorial key. The problem rather comes in terms of the calculator being able to store the products of the multiplication of several factorials. It is for that reason that some authors prefer to compute the Fisher's exact test by transforming the factorials into a log and then multiplying them. However, since my assumption is that you will have a computer handy, I prefer to discuss the simple definitional formula of the test.

Formula 8.5

Formula for computing the exact probability of observing a table with cells *a*, *b*, *c*, *d* (Fisher's exact test).

$$p(a, b, c, d) = \frac{(a+b)!^*\,(c+d)!^*\,(a+c)!^*\,(b+d)!}{n!^*a!^*b!^*c!^*d!}$$

Where a, b, c, and d are defined in Table 8.3.

Let us consider a case in which an anthropologist is determining if two different families in the village he is studying own a significantly different number of sheep and goats. The data he collected are:

Observed data			
Type of animal	Family 1	Family 2	Totals
Sheep	2	4	$a + b = 6$
Goats	3	2	$c + d = 5$
	$a + c = 5$	$b + d = 6$	$n = 11$

The exact probability of obtaining this table is:

$$p = \frac{6!5!5!6!}{11!2!4!3!2!} = \frac{720^*120^*120^*720}{39{,}916{,}800^*2^*24^*6^*2} = \frac{7{,}464{,}960{,}000}{22{,}992{,}076{,}800}$$

$$p = 0.3247.$$

Now we compute the probability of obtaining the next "worse" table, which we will call first "worse" table. This table is obtained by subtracting one from the diagonal with the lower numbers and adding one to the other diagonal, so that the row and column totals remain unchanged.

First "worse" table			
Type of animal	Family 1	Family 2	Totals
Sheep	1	5	$a+b=6$
Goats	4	1	$c+d=5$
	$a+c=5$	$b+d=6$	$n=11$

The exact probability of obtaining this table is:

$$p = \frac{6!5!5!6!}{11!1!5!4!1!} = \frac{720{*}120{*}120{*}720}{39{,}916{,}800{*}1{*}120{*}24{*}1} = \frac{7{,}464{,}960{,}000}{39{,}916{,}800{*}2880} = 0.0649.$$

Now we compute the probability of obtaining the next "worse" table, which we will call second "worse" table. This table is obtained by subtracting one from the diagonal with the lower numbers and adding one to the other diagonal of the first "worse" table. You will notice that the row and column totals remain unchanged.

Second "worse" table			
Type of animal	Family 1	Family 2	Totals
Sheep	0	6	$a+b=6$
Goats	5	0	$c+d=5$
	$a+c=5$	$b+d=6$	$n=11$

The exact probability of obtaining this table is:

$$p = \frac{6!5!5!6!}{11!0!6!5!0!} = \frac{720{*}120{*}120{*}720}{39{,}916{,}800{*}1{*}720{*}120{*}1} = \frac{7{,}464{,}960{,}000}{39{,}916{,}800{*}86400}$$
$$= 0.002164.$$

Now that we have reached a "worse" table with a zero frequency, we can compute the probability of the observed table and all "worse" outcomes *in one direction* as: $p = 0.3247 + 0.0649 + 0.0022 = 0.3918$.

Luckily for those of us who enjoy spending time with our calculators working out factorials, to compute a two-tailed Fisher's test we need to obtain the probability associated with the observed table and all worse outcomes in one direction *and then* the probability associated with outcomes in the other tail. Everything we have done up to now has been in one direction: we started with family one having two sheep, then we created a new table in which family one had one sheep and we stopped with the table in which family one had zero sheep. As you can see, we only moved in one direction. Only if the sum of row one is the same as the sum of row two can we assume that the probability of a two-tailed test is twice that computed in a one-tailed test. We need to compute the probability associated with the other tail. When we have it, we add it to the probability on one tail already computed to obtain the two-tailed probability. There are two frequently used ways in which the computation of the second tail is explained.

1. If $a^*d - c^*b$ in the original table is negative then set to 0 either b or c (whichever is smaller) and adjust the other cells to maintain the row and column numbers. This is going to be the most extreme case. Compute the probability of this table, and increase b and c by 1 and decrease a and d by 1 to compute the next table and its probability. Continue until you get a probability that is greater than the probability associated with the observed data (which in our case was 0.3247). If on the contrary $a^*d - c^*b$ is positive then set to 0 either a or d and adjust the other cells to maintain the row and column numbers. Compute the probabilities as explained in this paragraph. In our example $a^*d = 2^*2 = 4$ and $c^*b = 3^*4 = 12$. Thus 4–12 = −8. Therefore we need to set c to 0, which is a smaller number (3) than is b (4).
2. Find the smallest row total and the smallest column total and identify the cell in which both of these totals intersect in the *most extreme table you formed* to compute the one-tailed probability. Then subtract from the row total the element in that table where both smallest row and column intercept. The result of this subtraction should be put in the place where the smallest row and column intercept in a new table, and the other elements of the table are written out so that the row and column totals are preserved. In our example the lowest row total is 5, and the lowest column total is also 5. This row and this column intersected at c, where the value of the cell in the "most extreme table" (the last one) is 5. If we subtract 5 from the row total we get $5 - 5 = 0$. We set c to 0 and adjust the rest of the table to maintain the same row and column totals.

You can see that both ways of starting the second tail of the Fisher's test lead us to set c to 0. This is going to be the most extreme table on the other tail. Compute the probability associated with this table. If the probability is greater than that of the original (observed) table then this table does not contribute at all to the two-tailed probability. If it is not, then construct another less-extreme table and compute its probability. You should stop when you obtain a probability higher than the probability of the original table.

We proceed to compute the probability associated with the other tail: our "worse" table is:

Type of animal	Family 1	Family 2	Totals
Sheep	5	1	$a + b = 6$
Goats	0	5	$c + d = 5$
Totals	$a + c = 5$	$b + d = 6$	$n = 11$

The probability of this table is:

$$p = \frac{6!*5!*5!*6!}{11!*5!*1!*0!*5!} = \frac{720*120*120*720}{39{,}916{,}800*120*1*1*120}$$
$$= \frac{7{,}464{,}960{,}000}{39{,}916{,}800*14,400} = \frac{7{,}464{,}960{,}000}{57{,}480{,}192} = 0.013.$$

Because this p value is not more than the probability of our first (raw) table, which was 0.3247, we continue.

Our next-to "worse" table is:			
Type of animal	Family 1	Family 2	Totals
Sheep	4	2	$a+b=6$
Goats	1	4	$c+d=5$
	$a+c=5$	$b+d=6$	$n=11$

The probability of this table is:

$$p = \frac{6!*5!*5!*6!}{11!*4!*2!*1!*4!} = \frac{720*120*120*720}{39{,}916{,}800*24*2*1*24}$$
$$= \frac{7{,}464{,}960{,}000}{39{,}916{,}800*1152} = \frac{7{,}464{,}960{,}000}{45{,}984{,}153{,}600} = 0.1623.$$

Because this p value is not more than the probability of our first (raw) table, which was 0.3247, we continue.

Our next-to-next "worse" table is:			
Type of animal	Family 1	Family 2	Totals
Sheep	3	3	$a+b=6$
Goats	2	3	$c+d=5$
	$a+c=5$	$b+d=6$	$n=11$

The probability of this table is:

$$p = \frac{6!*5!*5!*6!}{11!*3!*3!*2!*3!} = \frac{720*120*120*720}{39,916,800*6*6*2*6}$$
$$= \frac{7{,}464{,}960{,}000}{39{,}916{,}800*432} = \frac{7{,}464{,}960{,}000}{17{,}244{,}057{,}600} = 0.43.$$

Table 8.6 Do pairs of matched students improve their reading scores after completing two different reading programs (A and B)?

	Student outcome	Reading program B: Improve	Reading program B: Did not improve	Totals
Reading program A	Improve	5 (both improved)	7	12 students improved after taking program A
	Did not improve	10	2 (neither improved)	12 students did not improve after taking program A
		15 students improved after taking program B	9 students did not improve after taking program B	24 pairs

This probability is higher than that which we computed for the original table. Therefore, this table does not contribute to the two-tailed probability we want to estimate. We need to add the probabilities of the two tables on the second tail, namely, $p = 0.013 + 0.1623 = 0.1753$. Now we are able to estimate the two-tailed probability associated with our original table: it is the sum of the probabilities on both sides; thus: $p = 0.3918 + 0.1753 = 0.5671$.

There are several "warnings" I would like to give you about the Fisher's test: you are likely to find several formulae in different books on how to compute the test. Several authors prefer to work with logs of the factorials, which allow the analyst to store in a calculator's memory manageable numbers. Therefore, do not be surprised if you see a different formula being used. In addition, if you decide to use one of the many "online calculators" for computing Fisher's exact test, you should really understand what you are getting. I could only get them to give me the two-side total probability (of the original table and all worse cases). It may be the case that you want to get the exact probability of the table you entered and not the probability of it and all worse cases. SAS computes the exact probability of the table, the left tail and the right tail probabilities and the two-tailed probability.

8.7 The McNemar test for a matched design

At first it may seem strange to speak of a matched design when we are talking about frequencies. However, let us take a look at the example shown in Table 8.6: you can match 24 pairs of students of the same gender, age, grade, and reading score in a standardized test and expose each member of the pair to two different reading programs which we will call A and B. The question we are asking here is: since the students have been matched for their pre-program reading score and for age, grade, and gender, will they respond differently (by improving or not improving their reading scores) after having being exposed to the two different reading programs? After a week you can re-test the students to determine how many of them improved and how many of them did not improve their reading scores.

The table shows that there were five pairs whose members did improve while there were two pairs who did not improve. Because the members of these pairs experienced the same outcome, these members belong to a **concordant pair**. In contrast, a total of 17 pairs are **discordant pairs:** for ten of these pairs the student exposed to reading program A did not improve while the student exposed to reading program B did improve and for seven pairs the student exposed to reading program A did improve while the student exposed to reading program B did not. As you can imagine, our interest centers on the discordant pairs since they are going to provide us with information about the effectiveness of the treatment effect: do more students improve when exposed to reading program A (7 students of the discordant pairs improved when exposed to A) rather than B (10 students of the discordant pairs improved when exposed to B)?

Formula 8.6

Formula for computing the McNemar's test with Edward's correction.

$$X^2 = \frac{(|b - c| - 1)^2}{b + c}, \quad \text{with df} = 1.$$

Where b and c are defined in Table 8.3. The Edward's correction refers to the "-1" which is subtracted from the difference between the values of b and c, and it makes the test more conservative.

Let us compute the McNemar test to test the hypothesis that improvement in reading scores is independent of reading program.

$$X^2 = \frac{(|7 - 10| - 1)^2}{7 + 10} = 0.235.$$

Because the X^2 is under the value of 1, we do not even bother to check if this is significant. The two reading programs did not differ significantly in their success.

8.8 Tests of goodness-of-fit and independence of variables. Conclusion

When should you use the G, Kolmogorov–Smirnov, Fisher's, the McNemar, or the chi-square test? The Kolmogorov–Smirnov and the McNemar tests are geared to specific situations so that it should be pretty obvious when they are appropriate: if your count data have an order you should use the former, and if they are recorded in matched pairs you should use the latter.

Concerning the other designs, Fisher's exact test is the choice test for a 2×2 table with expected frequencies under 5. Concerning the *G* and the X^2 tests, although they consistently produce very similar statistics (and therefore usually lead to the same decision regarding the null hypothesis) the G test is certainly drawing more and more adherents. This is a trend that perhaps should begin to include more and more anthropologists, who seem to love their chi-square. There is no question however that the two tests provide essentially the same results. For example, I performed each of the tests in this chapter with both tests and I achieved basically the same result whether I was using the G or the chi-square test. Even the Williams' correction did not change the result much, while Yates' correction did tend to reduce the test's power exceedingly (Sokal and Rohlf 1995). However, an issue to consider is what is available in the computer package you are using. You might wish to apply the Williams' correction only to find out that your computer package does not compute it! My general advice is: use Yates' correction only when you do not have access to a computer program. Otherwise use Fisher's exact test or Williams' correction. It seems that the G test is the way of the future, just as the chi-square test is on its way out to retirement.

8.9 The odds ratio (OR): measuring the degree of the association between two discrete variables

With the chi-square and the G tests we have been able to determine if two variables have a statistically significant association (in our example above, years of formal education in mothers was not independent of their using traditional medicine). The **odds ratio** allows us to look deeper into the nature of the relation between the two variables so that we can make statements about the probability associated with one event given another one. For example, we can ask what the odds are that a mother with high levels of education will not use traditional medicine in comparison with a mother who has low levels of education.

It never hurts to remind ourselves of the difference between the probability and the odds of an event. What is the probability that I will get a 6 if I roll a fair dice? The probability is $p = \frac{1}{6}$. The odds of an event is *not* the same thing as the probability of the event. Instead the **odds** is the frequency of being in one category relative to the frequency of not being in that category. Odds are stated in terms of one category of a variable relative to the sum of all the *remaining* categories. If there are 6 categories (such as in a dice), the odds of getting any one number are 1 to 5. The **conditional odds** are the chances of being in one category of a variable relative to the remaining categories of that variable, given a specific category of a second variable (for example, the probability of using traditional medicine given that a mother has low educational level). The **odds ratio (OR)** compares two conditional odds, in a single statistic, which is formed by dividing one conditional odds by another. If an odds ratio includes the value of one in its confidence interval, then the ratio is not statistically significant.

Before we apply the computational formula for obtaining the odds ratio (formula 8.7), let us investigate how we got to it. The odds ratio can be understood as follows (please refer to Table 8.3):

$$OR = \frac{\left[\frac{a}{(a+c)}\right] \Big/ \left[\frac{c}{(a+c)}\right]}{\left[\frac{b}{(b+d)}\right] \Big/ \left[\frac{d}{(b+d)}\right]}$$

$$OR = \frac{\left[\frac{\textit{exposed and positive}}{\textit{all exposed}}\right] \Big/ \left[\frac{\textit{exposed and negative}}{\textit{all exposed}}\right]}{\left[\frac{\textit{unexposed and positive}}{\textit{all unexposed}}\right] \Big/ \left[\frac{\textit{unexposed and negative}}{\textit{all unexposed}}\right]}$$

This formula, written in terms of exposure to a risk factor and in terms of developing the disease clarifies the meaning of the odds ratio. In a nutshell we have two ratios in one: in the numerator we are saying "... out of all those exposed, let us take a ratio of those who developed the disease over those who did not develop it." In the denominator we are saying "... out of those who were not exposed, let us take a ratio of those who developed the disease over those who did not develop it." The odds ratio allows us to have all of this information in one single number: it tells us what the odds are of someone developing the disease after exposure to the risk factor to the odds of someone developing the disease without being exposed to the risk factor. This is best explained with an example.

Let us use the data on mothers' years of formal education and their use of traditional medicine. However, let us now consider only mothers with high and low levels of education so that each variable has two levels. In this example variable A is use of traditional medicine and variable B is education.

	Education level		
Use of traditional medicine	High	Low	Total
No	14	8	22
Yes	3	14	17
Total	17	22	$n = 39$

Let us say that previous work indicates that more formally educated women are less likely to use traditional medicine. The odds ratio is going to allow us to determine what the odds are that a mother with high levels of education will not use traditional medicine in comparison with a mother with low levels of education. If the OR is significantly greater than one, then the odds that a woman with high education will not use traditional medicine is significant. If the OR is significantly lower than one, then the odds that a woman with high education will use traditional medicine is significant. If the OR is not significantly different from 0, then women with low and high levels of education are equally as likely to use traditional medicine.

Formula 8.7

Formula for the odds ratio.

$$OR = \frac{a^*d}{c^*b}$$

If any of the cells has frequencies of 0 then

$$OR' = \frac{(a + 0.5)^*(d + 0.5)}{(c + 0.5)^*(b + 0.5)}$$

Where a, b, c, and d are as defined in Table 8.3.

Let us compute the odds ratio for our education/traditional medicine example. What are the odds that a highly educated woman will not use traditional medicine in contrast with less-educated women? $OR = \frac{14^*14}{3^*8} = \frac{196}{24} = 8.166$.

The odds that highly educated women will not use traditional medicine when compared with less-educated women is much higher than an odds of 1. A highly educated woman has 8.166 odds of not of using traditional medicine when compared with less-educated women. You can of course look at the same question from the perspective of women with lower levels of formal education. We are going to ask the same question, that is, in terms of the women NOT using traditional medicine: what are the odds that a less formally educated woman will NOT use traditional medicine in comparison with a highly educated woman? In this case the $OR = \frac{3^*8}{14^*14} = \frac{24}{196} = 0.122$. Less-educated women are 0.122 times as likely NOT to use traditional medicine than highly educated women (if this is difficult to wrap your mind around, you need to consider that the OR in this case is under 1. This means that the odds are very low that women with low levels of formal education will *not* use traditional medicine). The reason I wanted to compute both odds ratios is that some computer programs will insist on computing the odds ratio looking at "one side of the coin" only, even if you change the way in which you enter your data!

Is the 8.166 OR significant? To answer that question we need to look at the confidence intervals of the odds ratio. However, here we have a problem: the distribution of the odds ratio is highly skewed since the odds ratio cannot take a negative value. For this reason, it is customary to transform the odds ratio into its natural log, which does have an approximately normal distribution and allows us to compute confidence intervals. Quite simply, if the confidence interval includes the value of 1 then the odds are not significant. This should be familiar to you given our discussion of the G test: if we divide a number by a very similar number then we will get something very similar to 1. The more different the two numbers, the more different the quotient will be from 1.

Formula 8.8

Formula for the computation of the 95% confidence interval (CI) for the natural logarithm of the odds ratio.

$$95\%\mathrm{CI(OR)} = \ln[ln\,(OR) \pm 1.96^*se\,[(OR)]]$$

where the standard error of the odds ratio is: $se\,(OR) = \sqrt{\frac{1}{a} + \frac{1}{b} + \frac{1}{c} + \frac{1}{d}}$
If any of the cells in the table have frequencies of 0 then

$$se\,(OR) = \sqrt{\frac{1}{a+0.5} + \frac{1}{b+0.5} + \frac{1}{c+0.5} + \frac{1}{d+0.5}}$$

Please note that you have to: (1) Transform the OR into its natural log, and that (2) Once you obtain the limits of the interval you transform both limits into their natural logs as well.

Let us compute the confidence intervals for the OR computed above ($OR = 8.166$). We had established that a highly educated woman is 8.166 as likely not to use traditional medicine as is a less-educated woman. To compute the 95% confidence intervals for this OR we first need to compute the standard error of the OR: $se = \left(\sqrt{\frac{1}{14} + \frac{1}{8} + \frac{1}{3} + \frac{1}{14}}\right) = \sqrt{0.60} = 0.775$. We also need to compute the ln of the OR which is ln(8.166) = 2.09998. Therefore the 95%CI (OR) $= \ln\left[ln\,(OR) \pm 1.96^{*}se\,[ln\,(OR)]\right] = \ln\left[2.09998 \pm 1.96^{*}0.774\right] = \ln\left[0.57\right], \ln\left[3.6\right] = 1.79{,}37.22$. The confidence intervals do not include one. Therefore, the odds of a woman who is highly educated NOT using traditional medicine in comparison with a woman who is less educated is significantly different from one.

8.10 The relative risk (RR): measuring the degree of the association between two discrete variables

The relative risk (RR) is another measure which evaluates the strength of the association between two discrete variables, whose "meaning" is quite similar to that of the odds ratio. The confidence intervals of the RR are less frequently computed because the sampling distribution of the relative risk is less well understood, although SAS does compute them. The RR may come in handy if you have limited access to computer resources, and wish to have a quick assessment of the strength of the relation between the two variables. Let us look at the RR considering the two variables (A and B) shown in Table 8.3.

Formula 8.9

Formula for computing the relative risk (RR).

$$\text{RR} = \frac{a/(a+c)}{b/(b+c)}$$

Let us re-write the formula for the RR in terms of exposure to a risk factor and of developing the disease.

$$RR = \frac{{exposed\ and\ positive}/{all\ exposed}}{{unexposed\ and\ positive}/{all\ unexposed}}$$

Therefore, the relative risk is the probability of being positive for the disease among exposed individuals divided by the probability of being positive for the disease among unexposed individuals.

Practice problem 8.5

Let us practice the computation of the OR and the RR with some fictitious data on the probability of being positive for hypertension among migrants and non-migrants. Here the disease is hypertension, and the risk factor is being a migrant. The data are:

	Migrant	Non-migrant	Totals
Hypertense	48	12	60
Non-hypertense	72	103	175
Totals	120	115	$n = 235$

The odds ratio is: $OR = \frac{48*103}{72*12} = 5.722$. The odds of a migrant developing hypertension are 5.7 the odds of a non-migrant developing the disease. The relative risk for migrants to have hypertension is: $RR = \frac{48/120}{12/115} = 3.833$.

The analysis of frequencies in anthropology

Anthropologists are interested in the cross-cultural variation of violent encounters within and without a specific culture. Questions of interest are the prevalence of violence in different groups (and the causes of such differences) and variation in prevalence between the genders within groups. Scott and Buckley (2010) investigate evidence of interpersonal violence in two skeletal populations in the Nebira site in Papua New Guinea (1230–1650) and in the Taumako site in Salomon islands (1530–1698). The authors choose to examine the frequencies of trauma in both groups and between males and females within each group with a Fisher's exact test because of their small sample size. Scott and Buckley conclude that the evidence for trauma is significantly higher in Nebira rather than Taumako males, and that there was no difference in the frequency of trauma between males and females in both sites. The Fisher's exact test allowed Scott and Buckley to test null hypotheses of wide interest to anthropologists using a very small sample size (Scott and Buckley 2010).

Can formal education change young people's attitudes about acceptable marriage partners? Kisioglu *et al.* (2010) report results of a project in which students in Turkey were matched and members of each pair were exposed to different educational programs about consanguineous marriage. Kisioglu *et al.* report a significant difference between members of pairs exposed to different programs. Specifically, a McNemar's test was applied to 2×2 tables in which both variables had two levels (would you consider consanguineous marriage – yes or no – before the test and after the program). A McNemar test was appropriate in this case because the students had been matched before the test was given (Kisioglu *et al.* 2010).

According to Kohrt and Worthman (2009) anxiety disorders are 1.5 to 2.0 times more common in females than in males throughout the world. Why this is the case is a question ripe for anthropological investigation, one that requires cross-cultural investigation on the factors that affect the expression of anxiety in both genders. To this end, Kohrt and Worthman (2009) investigate the prevalence of anxiety in males and females and the factors which modulate anxiety in Tibet. The authors report that in males the odds that a subject with low social support will suffer from anxiety are 3.5 times greater (95% CI = 1.4–10.7) relative to subjects reporting high social support. However, in females there is no difference in the odds that a subject will suffer from anxiety between women of low and high social support (Kohrt and Worthman 2009).

8.11 Chapter 8 key concepts

- What are the null hypotheses of chi-square (for both goodness-of-fit and independence-of-variable tests)?
- What is the correct notation for the test statistic of a chi-square test?
- Why is the X^2 not parametric?
- Refer to the computational formula of X^2 and of the G test. Why is the deviation between the observed and expected frequencies computed the way it is?
- Yates' correction for continuity.
- Williams' correction.
- When should you use the X^2, the Fisher's exact test, McNemar, and the Kolmogorov–Smirnov tests? What corrections are appropriate and when?
- Explain the odds ratio.
- Explain the relative risk.

8.12 Computer resources

1. Neither PASW nor SAS is terribly user-friendly when it comes to analyzing frequencies. PASW requires you to "weight" the data by the observed frequencies (an extra step) and it does not allow you to enter columns of character data such as "medium,

high, low." SAS allows you to enter character data but it does not produce the G test for a simple goodness-of-fit test.

2. As far as I can tell neither SAS nor PASW perform the Williams' correction.
3. I would not take it for granted that every single computer package computes the odds ratio and its natural log.
4. While PASW allows you to obtain the odds ratio using its menus, SAS requires that you write code for obtaining it. PASW does not print the standard error of the OR, while SAS does.
5. Please note that both PASW and SAS print the odds ratio un-transformed (say 8.166 above) while the confidence intervals are those computed using the log transform (say 1.79, 37.22 above).
6. For the McNemar test I don't have much good news: SAS does not apply the Edwards' correction, so its X^2 statistic will differ from that computed following the formula presented here in a manner that can be easily checked. In the case of our example on the two reading programs SAS computes the following: $X^2 = \frac{(|7-10|)^2}{7+10} = 0.5294$. PASW does not print the X^2 statistic but only the p value, which unfortunately never coincides with that printed by SAS. So, you don't really know what PASW is doing.
7. I could find no way of doing the Kolmogorov–Smirnov goodness-of-fit test in SAS. PASW does compute it but it requires the user to "weight" the data by the observed frequencies.

8.13 Chapter 8 exercises

1. An anthropologist working with a horticultural group wishes to determine if the group's meat consumption differs seasonally. The researcher records the number of large animals consumed (shared) by the entire community during the rainy and dry seasons. Test the hypothesis that the frequency of animals consumed does not differ seasonally. Perform a chi-square and a G test.

	Number of animals
Season	
Dry	40
Wet	70
Total	110

2. An anthropologist working with a horticultural group notices that when the community shares the meat of a large animal, the hunter gives as a present the head of the animal to somebody of his choice. Both males and females of high and low status receive the present. The anthropologist records the following data set. Use it

to test the hypothesis that gender and status of recipients are independent. Perform a chi-square test and G test. What are the odds that a female of high status will receive the head in contrast with a male of high status?

	Gender		
Status	Female	Male	Total
High	17	20	37
Low	18	9	27
Total	35	29	64

3. The following data were collected during a paired-design project in which the McNemar test is appropriate. Twenty-three families sent one of their sons to study at the university while keeping another son at the family farm to help run it and eventually take it over. After both sons are married, an anthropologist measures their wealth in terms of number of horses owned by each son. The anthropologist decides to measure wealth at two levels: high number of horses and low number of horses. In this project the pair is the two sons and the two variables whose independence we are testing are: (1) Education (university or stayed at home) and (2) Number of horses (high or low). Five pairs were concordant for high numbers of horses while six were concordant for low number of horses. Did the treatment have an effect on wealth accumulation?

		Son attended university		
Son stayed at the farm	Number of horses	High	Low	Total
	High	5	7	12
	Low	5	6	11
	Total	10	13	23

9 Correlation analysis

Correlation analysis is a very popular statistical technique whose purpose is to determine if two variables co-vary. What is peculiar about correlation analysis is that neither variable is dependent or independent. What we want to determine is whether there is a statistically significant relation between both variables. As you will learn in the regression chapters, in regression analysis we designate the independent variable with an X and the dependent variable with a Y. To clarify that in correlation analysis we do not have such variables we will designate one variable as Y_1 and another one as Y_2. With this notation we imply that both variables in correlation are free to vary and are out of the control of the researcher.

When we apply the parametric correlation test, our variables must be continuous or discontinuous numeric. If the variables are discontinuous numeric then they should have a broad enough range, and not have few values (such as 0 and 1 or 1–5). When we apply the non-parametric correlation tests, we have more freedom in terms of the type of data we can analyze with the test. For example, we can use data that are approximations (estimates) or that are not well measured ("five or more"). Whatever type of data we are analyzing, what must be clear is that if we find that there is a significant correlation between two variables, we are not in any way saying that one variable causes another one. I am sure you have heard the saying "Correlation does not mean causation." This is so true and so frequently forgotten! The natural and social world is full of **spurious correlations**, correlations which arise only because of chance and which have no meaning or importance in the natural and social world.

9.1 The Pearson product-moment correlation

The Pearson correlation is a commonly applied parametric test which quantifies the relation between two numeric variables, and tests the null hypothesis that such relation is not statistically significant. The correlation between the variables is quantified with a coefficient whose statistical symbol is r, and whose parametric symbol is ρ ("rho"). The coefficient ranges in value from -1 to $+1$. If r is negative, then as Y_1 increases, Y_2 decreases (Figure 9.1). If r is positive then as Y_1 increases, Y_2 increases as well (Figure 9.2). If r is not statistically significantly different from 0, then there is no significant relation between Y_1 and Y_2 (Figure 9.3). Thus, in correlation analysis the *null hypothesis* is that the parametric correlation between the two variables is 0; the

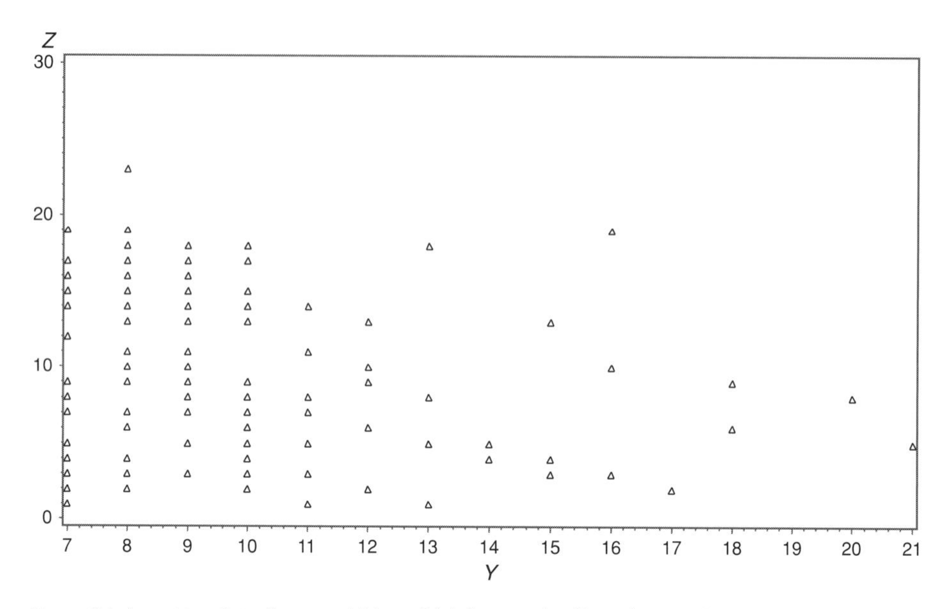

Figure 9.1 A scatter plot of two variables which have a significantly negative correlation.

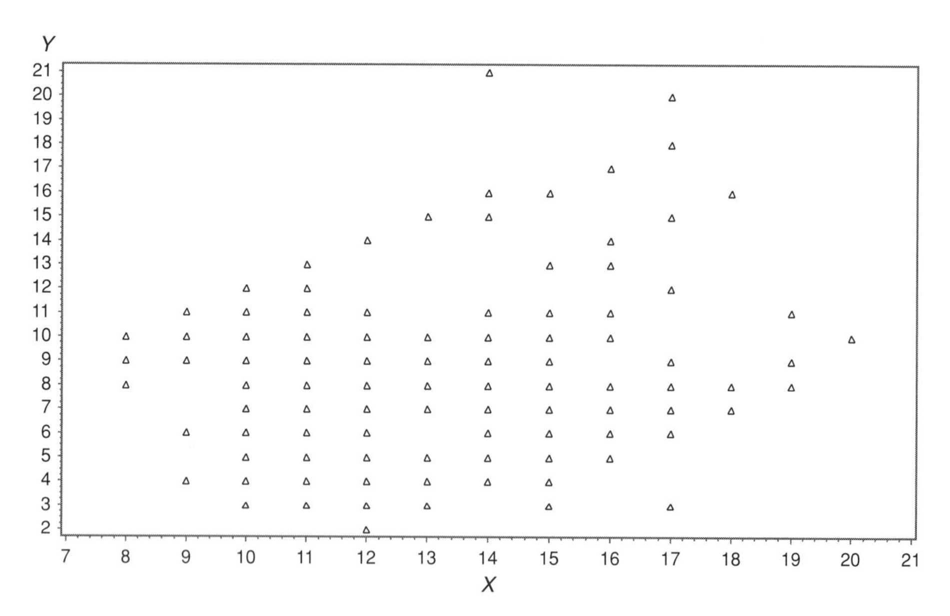

Figure 9.2 A scatter plot of two variables which have a significantly positive correlation.

usual two-tailed test *null hypothesis* is H0: $\rho = 0$. A one-tailed test is possible as well, although it should be used only when there are compelling reasons for it.

The reader is by now familiar with the fact that many statistical techniques assume a sample data set to be normally distributed. Indeed, for analysis of variance, it was stressed that every sample be tested for normality of distribution. Correlation analysis

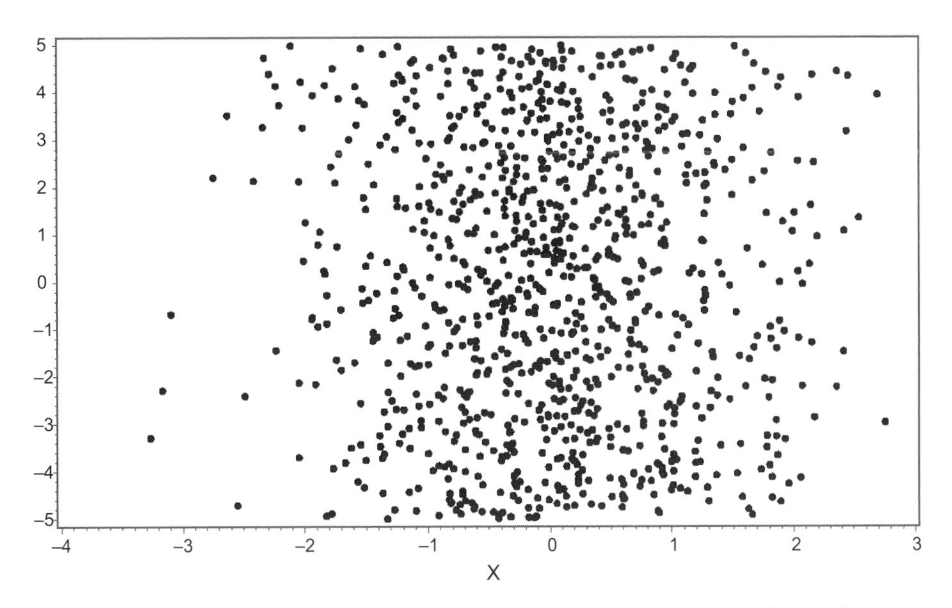

Figure 9.3 A scatter plot of two variables whose correlation does not differ significantly from 0.

also has a normality assumption, one that is more difficult to test: the data need to be **bivariate normal**. That is, the frequency of subjects with extreme measures *in both variables* should be that predicted by the rules of the normal probability distribution. Thus, a bivariate distribution looks like a three-dimensional normal curve (a "fat" normal curve) in which both variables have a normal distribution. Let us try to think of two standard normal distributions (that is, both variables have a mean of 0 and a standard deviation of 1). Let us plot both variables on a piece of paper centered on the value of 0, and let us look at the array of points. Figure 9.3 shows such a distribution for 1000 data points. If we could truly "stack" these points, that is, if the points were not flat dots on a piece of paper but if they were marbles, we would be looking at the top of a volcano-shaped array of marbles. The reason we see an oval of points is that the plot is not three-dimensional and that both distributions are centered on the value of 0. You can see that there are very few observations that depart sharply from the scatter of points, indicating that most of them fall close to the value of 0, which is the mean for both distributions. Therefore we can assume that these observations follow a bivariate normal distribution.

Although we do not have a simple test of normality for the bivariate normal distribution, we can at the very least do a plot of the observations, and inspect the array of points to see if we have obvious departures from bivariate-normality. Such departures include not only extreme observations away from the main array of points, but also an obvious curved relation between the Y_1 and the Y_2 and one in which the variance of Y_2 increases as the value of Y_1 increases (so that the plot looks like a fan). As usual, things are easier said than done: the plot of two unrelated random and bivariate normal variables shown in Figure 9.3, for example, was achieved by plotting 1000 observations. If the data set consisted of a few observations it would not be possible to ascertain if the

Table 9.1 Selected critical values of the Pearson correlation coefficient.

Degrees of freedom = $n - 2$	Levels of significance for a two-tailed test		
	0.05	0.02	0.01
9	0.602	0.685	0.735
10	0.576	0.658	0.708
15	0.482	0.558	0.606
20	0.423	0.492	0.537
25	0.381	0.445	0.487
30	0.349	0.409	0.449
50	0.273	0.322	0.354

bivariate normal assumption had been violated or if a curved relation is present. In these last few paragraphs I have been alluding to another Pearson correlation assumption: *the data should be linearly related*. Therefore, the Pearson correlation should not be applied to data which have a curved relation. If the variables have a curved relation, then a transformation of one or both variables should be attempted. Thus, a plot of the data should be the first step in correlation analysis. In this step we would be asking if there are any obvious departures from bivariate normality and from the assumption of linearity.

Mathematically speaking, the Pearson coefficient is very easily computed. In fact, the reader is familiar with all the components needed for the computation of r. Although the coefficient may be transformed into a t score, and a t table may be used for significance testing, the coefficient is more conveniently compared with a table of critical values of r, where the critical value is associated with $\text{df} = n - 2$ (Table 9.1). As usual, if the coefficient computed with the data is greater than or equal to the critical value, then the null hypothesis of no correlation is rejected. The formula for computing the Pearson correlation is shown in formula 9.1.

Formula 9.1

The Pearson correlation coefficient.

$$r = \frac{SP_{Y_1Y_2}}{\sqrt{(SS_{Y_1})\,(SS_{Y_2})}}, \quad \text{where}$$

$$SP_{Y_1Y_2} = \sum Y_1Y_2 - \frac{(\sum Y_1)(\sum Y_2)}{n}$$ is the **cross-products sum of squares**

$$SS_{Y_1} \quad \text{and} \quad SS_{Y_2} = \sum Y^2 - \frac{(\sum Y)^2}{n}$$, with the appropriate subscripts.

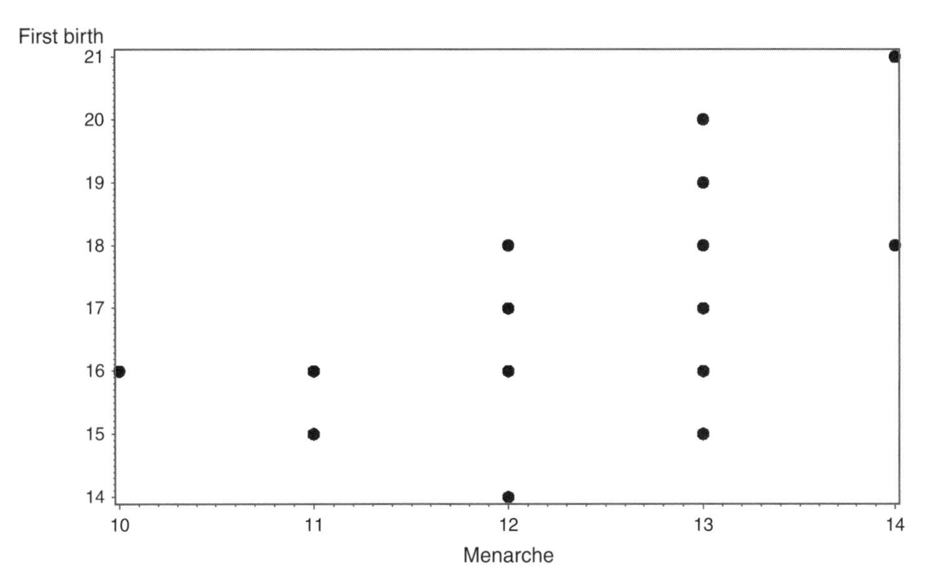

Figure 9.4 A scatter plot of the age at menarche and the age at first birth.

You can see that the formula of the correlation coefficient is basically a measure of how the two variables co-vary (in the numerator) divided by some measure of the variation within each of the two variables (in the denominator). The distribution of correlation coefficients (for a specific sample size) is normal with a mean of 0. We now proceed with an example on the computation of the correlation coefficient.

Let us presume that an anthropologist is interested in finding out if there is a significant correlation between a group of females' age at menarche, and the age at which they experience their first birth. The anthropologist does not have a dependent variable which needs to be explained in terms of an independent one. Perhaps both variables co-vary, instead of one causing the other one: it is possible that females with late and early menarche commence sexual behavior at comparable ages, but that those whose menarche occurs first are able to conceive first. Or it is possible that those females with early menarche do start sexual behavior before, and as a result conceive first. The interest of the researcher is simply to establish if both variables co-vary. Thus, she tests the null hypothesis that there is no correlation between the two variables: the data are in an Excel file called "menarche and first birth."

The first step is to plot the data, and check for presence of obvious departures from a bivariate normal distribution. None is observed in Figure 9.4 (the reason the plot does not look as "smooth" as do Figures 9.1–9.3 is that that variables are not continuous and their range is rather limited).

We follow the usual steps when testing a hypothesis:

1. We state the null and alternative hypothesis: H0: $\rho = 0$, H1: $\rho \neq 0$.
2. The alpha level will be 0.05. The degrees of freedom are $df = n - 2 = 28$. For these degrees of freedom the critical value is not listed in Table 9.1. Therefore, we use the

more conservative values for df = 25 which is 0.381. If the correlation coefficient we compute is greater than or equal to 0.381 we reject the null hypothesis.

3. The sample has been collected and given to us.

Let us refer to the age at first birth as Y_1, and to the age at menarche as Y_2. We establish the following quantities:

$$n = 30, \quad \sum Y_1 = 512, \quad \sum Y_1^2 = 8834, \quad \sum Y_2 = 367, \quad \sum Y_2^2 = 4519,$$
$$\text{and} \quad \sum Y_1 Y_2 = 6299.$$

First we compute the cross-products sums of squares: $SP_{Y_1}Y_2 = 6299 - \frac{(367)(512)}{30} = 35.53$.

The sums of squares are:
$$SS_{Y_1} = 8834 - \frac{(512)^2}{30} = 95.87.$$
$$SS_{Y_2} = 4519 - \frac{(367)^2}{30} 29.37.$$

Therefore: $r = \frac{35.53}{\sqrt{(95.87)(29.37)}} = 0.67$. With degrees of freedom df = 25, our correlation coefficient is significant at the 0.01 level. Our conclusion is that both variables are highly correlated in this community.

Practice problem 9.1

The data set called "infant birth weight" has information obtained during the pregnancy of 55 women such as the baby's birth weight, the mother's pregnancy weight, the maternal blood pressure at last check, and whether the mother smoked during pregnancy. Is there a correlation between mother's weight gain and mother's systolic blood pressure? We follow the usual steps when testing a hypothesis:

1. We state the null and alternative hypothesis: H0: $\rho = 0$, H1: $\rho \neq 0$.
2. The alpha level will be 0.05. The degrees of freedom are df = $n - 2 = 53$. For these degrees of freedom the critical value is not listed in Table 9.1. Therefore, we use the more conservative values for df = 50 which is 0.273.
3. The sample has been collected and given to us.

A plot of the data (not shown) indicated that the relation between both variables is linear and in compliance with a bivariate distribution. Let us call mother's birth weight Y_1 and mother's systolic blood pressure Y_2. The necessary statistics are:

$$n = 55, \quad \sum Y_1 = 773.9894, \quad \sum Y_1^2 = 13{,}895.05, \quad \sum Y_2 = 6750.125,$$
$$\sum Y_2^2 = 937{,}544.55, \quad \sum Y_1 Y_2 = 113{,}041.6487.$$

First we compute the sum of cross-products:

$$SP_{Y_1}Y_2 = 113{,}041.6487 - \frac{(773.9894)(6750.125)}{55} = 18{,}050.28.$$

The sums of squares are: $SS_{Y_1} = 13,895.05 - \frac{(773.9894)^2}{55} = 3003.06$. $SS_{Y_2} = 937,544.55 - \frac{(6,750.125)^2}{55} = 109,104.78$.

Therefore: $r = \frac{18,050.28}{\sqrt{(3003.06)(109,104.78)}} = 0.997$. With degrees of freedom df $= 50$ (Table 9.1), our correlation is significant at $p \ll 0.01$.

9.2 Non-parametric tests of correlation

Anthropologists frequently ask correlation-type questions about a data set which does not lend itself to a Pearson correlation analysis because the data are ranks. A prestige scale of adults in a community comes to mind: If ten adult males are ranked in terms of their prestige, the ranker does not imply that the difference between individuals one and two is the same as the difference between individuals eight and nine and nine and ten. Indeed, many types of ranked data (reliance on domesticated animals, importance of cash in households' wealth, complexity of decoration, etc.) cannot be analyzed with a Pearson correlation for this precise reason. In this section we cover two non-parametric tests of correlation: the Spearman test (r_S) and the Kendall's coefficient of rank (τ). The former is more broadly used in part because it is easier to compute by hand. However, with the current ease of access to computers the analyst does not need to be concerned about spending too many hours with pen and pencil in hand trying to compute Kendall's test. On the other hand, Kendall's test may be applied to data that violate an assumption of Spearman's test, so it may come in handy during difficult times!

9.2.1 The Spearman correlation coefficient r_S

It is important to stress that the Spearman correlation coefficient should be denoted as r_S, that is, with an "s" subscript, in order to differentiate it from the Pearson correlation coefficient. Although it is non-parametric, the Spearman correlation does have the assumption that the data be related in a linear manner. Therefore, the data should be plotted and inspected for violations to this assumption. We may apply it to raw data which were collected as ranks or to data that are transformed into ranks, where the ranking is done in the same manner as in the non-parametric tests chapter. The test proceeds as follows (after the inspection of the plot):

1. Rank both variables independently of each other, assigning ties if necessary. The rest of the test works with the two columns of ranks. Frequently the raw data will be ranks.
2. Create two columns of numbers. The first one should contain the ranks of the first variable rank-ordered so that if observations number 3, 4, and 6 had the first, second, and third rank, they are ranked first, second, and third in the first column. The second column should have the ranks of the other variable, not ranked, but written out so that the observations have both variates next to each other. For example, if observations

number 3, 4, and 6 were ranked 2nd, 1st, and 3rd for the 2nd variable, then their ranks in the second column will be 2nd, 1st, and 3rd, even though their ranks in the first column were 1st, 2nd, and 3rd. Therefore, one of the columns will be ranked; the other one may or may not be. The *null hypothesis* states that there is no association between the two columns of ranks. If there is total agreement (that is, if the observations are ranked in the same manner in both columns), then the null hypothesis will be rejected. If there is no agreement, the null hypothesis of no association will be accepted.
3. The difference (D) between the two ranks will be obtained and squared (D^2). The formula for the r_s coefficient is shown in formula 9.2.

Formula 9.2

The Spearman rank-order correlation coefficient.

$$r_s = 1 - \left[\frac{(6)\left(\sum D^2\right)}{(n)\left(n^2 - 1\right)}\right], \quad \mathrm{df} = n - 2,$$

where D is the difference between the two ranks. The r_s coefficient is compared with the table of critical values of the Pearson correlation coefficients.

Let us work first with a data set (called prestige and chickens) that includes both a quantitative measure and a set of ranks as its two variables. Let us presume that an anthropologist is interested in investigating if, in a particular culture, prestige among adult males is correlated with the number of chickens the males own. It is possible that chickens are not the "high prestige" farm animal, and that men who own lots of chickens do so because they cannot afford more expensive animals. Or it is possible that chickens are a good prestige marker, and that the higher the prestige, the higher the number of chickens. The researcher first interviews the religious specialist, who is said to be outside the prestige hierarchy of adult males. The anthropologist asks the informant to rank-order the adult male heads of households in the community. This variable consists of ranks only, where the highest ranked male receives the highest number (I know that this may be confusing, but to clarify the issue of positive and negative correlation, I will assign a "1" to the lowest prestige and an "11" to the highest prestige). The researcher then proceeds to interview each male, and asks him how many chickens he owns. The anthropologist confirms to the best of his abilities the reported number of chickens.

The first step is to plot the data and determine if there is a violation to the assumption of a linear relation between the variables. The data appear to be linearly related (Figure 9.5).

We rank-order the data according to one of the columns. Since prestige is already ranked, we can order the observations according to it. The number of chickens was ranked, assigning ties when necessary. The ranked data are shown below, as are the differences between the two ranks. We follow the usual steps when testing a hypothesis:

1. We state the null and alternative hypothesis: H0: the correlation between prestige and number of chickens is not significantly different from 0. Please note that because this is not a parametric test, we do not use parametric notation.

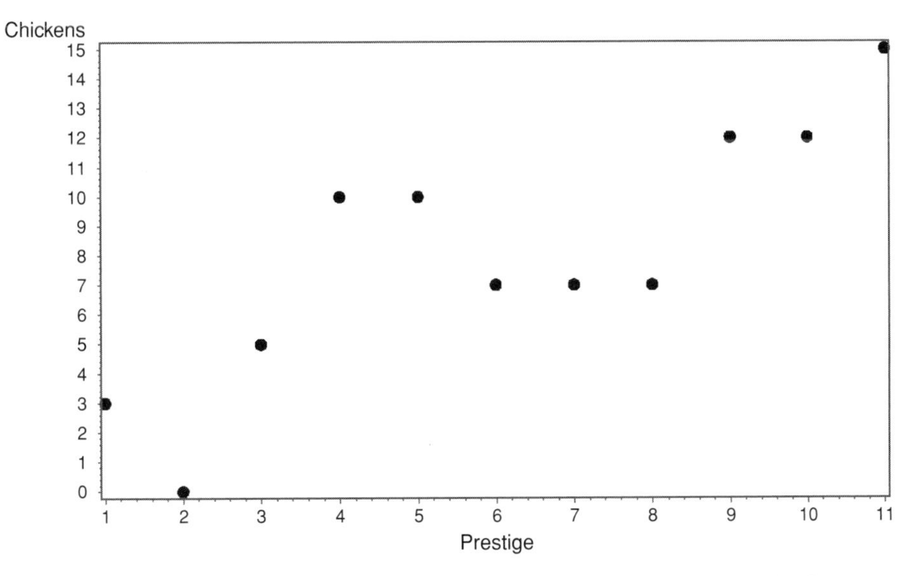

Figure 9.5 A scatter plot of prestige rank and the number of chickens owned by heads of households.

2. The alpha level will be 0.05. The degrees of freedom are df $= n - 2 = 9$. For these degrees of freedom the critical value is 0.602 (Table 9.1). If the correlation coefficient we compute is greater than or equal to the critical value we reject the null hypothesis.

ID	Prestige	Number of chickens	Rank of chickens	Difference between prestige and rank of chickens
3	1	3	2.0	−1.0
4	2	0	1.0	1.0
6	3	5	3.0	0.0
11	4	10	7.5	−3.5
8	5	10	7.5	−2.5
10	6	7	5.0	1.0
9	7	7	5.0	2.0
7	8	7	5.0	3.0
2	9	12	9.5	−0.5
5	10	12	9.5	0.5
1	11	15	11.0	0.0

We compute $\sum D^2 = -1^2 + \ldots 0^2 = 35$. $r_s = 1 - \left[\frac{(6)(35)}{(11)(11^2-1)}\right] = 1 - 0.16 = 0.84$, df $= 9$, $p < 0.01$. We reject the null hypothesis of no association between prestige and number of chickens with a high degree of confidence. As prestige increases so does the number of chickens.

Practice problem 9.2

Let us compute the Spearman correlation for the data on menarcheal and age at first pregnancy (the name of the data set is "Menarche and first birth"). Some of the data and their ranks are reproduced below. The null hypothesis is that menarcheal age and age at first pregnancy are not significantly correlated.

Age at menarche	Age at first birth	Rank for age at menarche	Rank for age at first birth	Difference between ranks	Difference squared
10	16	1.0	10.0	−9.0	81.00
11	15	4.5	4.0	0.5	0.25
11	16	4.5	10.0	−5.5	30.25
11	16	4.5	10.0	−5.5	30.25
11	15	4.5	4.0	0.5	0.25
11	15	4.5	4.0	0.5	0.25
11	15	4.5	4.0	0.5	0.25
12	18	13.0	22.5	−9.5	90.25
12	17	13.0	16.5	−3.5	12.25
12	18	13.0	22.5	−9.5	90.25
			⋮		
13	18	23.0	22.5	0.5	0.25
13	17	23.0	16.5	6.5	42.25
14	21	29.0	29.5	−0.5	0.25
14	21	29.0	29.5	−0.5	0.25
14	18	29.0	22.5	6.5	42.25

$$\sum D^2 = 1460.50.$$

$$\text{Therefore: } r_s = 1 - \left[\frac{(6)(1460.5)}{(30)(30^2 - 1)}\right] = 1 - \frac{8763}{26{,}970} = 0.675.$$

The correlation coefficient is compared with the critical value at 0.05 for df = 25 (0.381). We reject the null hypothesis. The two variables are significantly correlated.

9.2.2 Kendall's coefficient of rank correlation – tau (τ)

Although Kendall's coefficient is denoted by a Greek letter it does not refer to a parameter but to a test statistic which measures the strength of the association between two sets of ranks. Kendall's tau is quite useful when we need to compute a correlation between two sets of ranks which do not meet the assumption of linearity of the Spearman test. Please do not be surprised if you see not only different formulae for the computation of Kendall but also different suggestions on how to handle tied observations. Not surprisingly, computer packages also differ in the algorithm they use for their own computation, so it is possible that what SAS or PASW compute is not what you compute by hand. Indeed,

I found several desperate calls from analysts in the internet asking for help on how to solve the discrepant results obtained by hand and by computer packages. There is no question that the computation of Kendall's coefficient is rather involved if it is done by hand. I will demonstrate here what I consider to be the easiest way to compute it, which I prefer to present in two extreme cases: one with complete agreement and one with complete disagreement between the ranks.

Formula 9.3

Formula for the computation of Kendall's coefficient of rank correlation (τ).

$$\tau = \frac{N}{\sqrt{\left[n(n-1) - \sum T_1\right]\left[n(n-1) - \sum T_2\right]}}$$

$$\text{Where } N = 4^* \sum^{i} C - n(n-1),$$

$\sum^{i} C$ is the count of the ranks which are greater than the rank in column 2 at the current observation and

$\sum T_1$ is the sum of $t(t-1)$, where t is the number of ties which occurred in the first column to be ranked and

$\sum T_2$ is the sum of $t(t-1)$, where t is the number of ties which occurred in the second column to be ranked.

Let us compute Kendall's tau without any ties first. The first column is ranked from one to ten, while the second column is ranked from ten to one.

$i = 1 - 10$	Column 1	Column 2	Counts (C) of ranks in column 2 which are larger than the rank at current i
1	1	10	None (no rank in column 2 is larger than 10)
2	2	9	None (no rank in column 2 is larger than 9)
3	3	8	None
4	4	7	None
5	5	6	None
6	6	5	None
7	7	4	None
8	8	3	None
9	9	2	None
10	10	1	None
			$\sum C = 0$

To compute Kendall's tau we arrange the observations by sorting them by the first column of ranks, and writing down next to the first column the second column of ranks, as we did when we computed the Spearman rank correlation. To illustrate a positive and a negative correlation using Kendall's tau, let us agree that a rank of one is less than a rank of two and so forth. In this example we have ten individuals ($i = 1 - 10$) who were ranked one through ten for the first variable (let us say an estimated measure of wealth so that individual #1 has the lowest wealth while individual #ten has the highest wealth) and also ranked for prestige in the village (so that individual #10 has the highest prestige, individual #9 has the second highest prestige, and individual #1 has the lowest prestige rating). Thus, the two columns show exactly opposite ranks.

To compute $\sum^{i} C$ we ask how many observations in column two have ranks higher than the current rank. Since observation 1 has a rank of 10 in the second column there are no ranks higher than 10 in column two, so that its C (for count) is 0. This is the case for all other observations in column two so that the sum of the counts is 0.

$$\text{Thus N} = 4^* \sum^{i} C - n(n-1) = 4^*0 - 10\,(9) = -90, \quad \text{and}$$

$$\tau = \frac{-90}{\sqrt{[10^*\,(9) - 0]^*\,[10^*\,(9) - 0]}} = \frac{-90}{\sqrt{8100}} = \frac{-90}{90} = -1.$$

Kendall's tau is -1, indicating that the relation between both ranks is perfectly negative.

Let us now look at a data set which has perfect concordance between two ranks. In this example we have ten individuals ($i = 1 - 10$) who were ranked one through ten for the first variable (let us say an estimated measure of wealth so that individual #1 has the lowest wealth while individual #ten has the highest wealth) and also ranked for prestige in the village (so that individual #1 has the lowest prestige, and individual #10 had the highest prestige rating).

$i = 1 - 10$	Column 1	Column 2	Ranks in column 2 which are larger than the rank at current i	Counts
1	1	1	2 3 4 5 6 7 8 9 10	9
2	2	2	3 4 5 6 7 8 9 10	8
3	3	3	4 5 6 7 8 9 10	7
4	4	4	5 6 7 8 9 10	6
5	5	5	6 7 8 9 10	5
6	6	6	7 8 9 10	4
7	7	7	8 9 10	3
8	8	8	9 10	2
9	9	9	10	1
10	10	10	0	0
				$\sum = 45$

$$\text{Thus}\quad \mathrm{N} = 4^* \sum^{i} C - n(n-1) = 4^*45 - 10\,(9) = 90, \quad \text{and}$$

$$\tau = \frac{90}{\sqrt{[10^*\,(9) - 0]^*\,[10^*\,(9) - 0]}} = \frac{90}{\sqrt{8100}} = \frac{90}{90} = 1.$$

Kendall's tau is 1, indicating that the correlation between both columns of ranks is perfectly positive.

Let us now consider a situation in which the two columns of ranks have a positive relation and we have two individuals tied at the 3.5th level and two individuals tied at the 9.5th level. The data are:

$i = 1 - 10$	Column 1	Column 2	Ranks in column 2 which are larger than the rank at current i	Counts
1	1	1	2 3.5 3.5 5 6 7 8 9.5 9.5	9
2	2	2	3.5 3.5 5 6 7 8 9.5 9.5	8
3	3	3.5	5 6 7 8 9.5 9.5	**6.5***
4	4	3.5	5 6 7 8 9.5 9.5	6
5	5	5	6 7 8 9.5 9.5	5
6	6	6	7 8 9.5 9.5	4
7	7	7	8 9.5 9.5	3
8	8	8	9.5 9.5	2
9	9	9.5		**0.5***
10	10	9.5		0
				$\sum = 44$

* **Here is how we handle ties:** The first member of a pair of a tie contributes 0.5 to the sum of the counts but the second member of the pair does not. For example, the rank 3.5 contributes a count of 6.5 because there are six ranks higher than 3.5 and one which is equal to it following it. The second pair of rank 3.5 contributes a count of six because there are six ranks higher than 3.5.

Since we have ties we have to compute $\sum T_i = 2\,(2-1) + 2\,(2-1) = 4$, where 2 is the number of ties at each level. Please note that these ties are found only in the second column. For that reason this correction is only applied to the second element in the denominator.

$$\text{Thus}\quad \mathrm{N} = 4^* \sum^{i} C - n(n-1) = 4^*44 - 10\,(9) = 86, \quad \text{and}$$

$$\tau = \frac{86}{\sqrt{[10^*\,(9)]^*\,[10^*\,(9) - 4]}} = \frac{86}{87.97} = 0.977.$$

Finally let us now consider a situation in which the two columns of ranks have a negative relation and we have two individuals tied at the 3.5th level and two individuals tied at the 9.5th level. The data are:

$i = 1 - 10$	Column 1	Column 2	Ranks in column 2 which are larger than the rank at current i	Counts
1	1	9.5	None	**0.5***
2	2	9.5	None	0
3	3	8	None	0
4	4	7	None	0
5	5	6	None	0
6	6	5	None	0
7	7	3.5	None	**0.5***
8	8	3.5	None	0
9	9	2	None	0
10	10	1	None	0
				$\sum = 1$

***Here is how we handle ties:** The first member of a pair of a tie contributes 0.5 to the sum of the counts. For example, the first rank 9.5 contributes a count of 0.5 because it has a 9.5 after it. The second pair of rank 9.5 contributes a count of 0 because no other rank after it is equal to or greater than 9.5.

Since we have ties we have to compute $\sum T_i = 2(2 - 1) + 2(2 - 1) = 4$, where 2 is the number of ties at each level. Please note that these ties are found only in the second column. For that reason this correction is only applied to the second element in the denominator.

$$\text{Thus} \quad \text{N} = 4^* \sum^{i} C - n(n - 1) = 4^*1 - 10(9) = -86, \quad \text{and}$$

$$\tau = \frac{-86}{\sqrt{[10^* (9)]^* [10^* (9) - 4]}} = \frac{-86}{87.977} = -0.9775.$$

Practice problem 9.3

Let us compute Kendall's tau with a data set which consists of ranks only. A student is interested in evaluating the consistency of opinions held by two experts about archaeological societies' reliance on agriculture. The researcher asks two experts to rank 12 archaeological cultures on the societies' reliance on agriculture, with the most reliant society receiving the highest rank. The null hypothesis tested is that there is no association between the archaeologists' ranks.

Society	Rank 1	Rank 2	Ranks in column 2 which are larger than the rank at current *i*	Counts
2	1	3	4 9 8 6 12 5 7 11 10	9
7	2	4	9 8 6 12 5 7 11 10	8
1	3	1	9 2 8 6 12 5 7 11 10	9
12	4	9	12 11 10	3
5	5	2	8 6 12 5 7 11 10	7
8	6	8	12 11 10	3
11	7	6	12 7 11 10	4
9	8	12	none	
3	9	5	7 11 10	3
10	10	7	11 10	2
6	11	11	none	0
4	12	10	none	0
				$\sum = 48$

We follow the usual steps when testing a hypothesis:

1. We state the null and alternative hypothesis: H0: the correlation between both sets of ranks is not significantly different from 0.
2. The alpha level will be 0.05. The degrees of freedom are df $= n - 2 = 10$. For these degrees of freedom the critical value is 0.576 (Table 9.1).
3. The sample has been collected and given to us.
4. We compute Kendall's tau as follows:

$$N = 4^*48 - 12^*11 = 192 - 132 = 60$$

$$\tau = \frac{60}{\sqrt{[(12)(11)]^*[(12)(11)]}}$$

$$= \frac{60}{132} = 0.4545.$$ We do not reject the null hypothesis.

Correlation analysis in anthropology

To what extent inbreeding affects fitness is a question that requires cross-cultural data. In European societies few families have engaged in such extensive inbreeding and have been so thoroughly studied as has the Habsburg dynasty in Spain. During the Habsburg rule in Spain from 1516 through 1700 the family engaged in frequent consanguineous marriages between close relatives. The family's rule ended in 1700 with the death of Charles II, who suffered from numerous mental and physical disabilities. Alvarez *et al.* (2009) computed the inbreeding coefficient of Charles II and other members of the family and concluded that several Habsburg individuals had exceedingly high inbreeding coefficients, even higher than 0.20 (for comparison, the offspring of a first cousin mating has an inbreeding level of 0.0625). The authors computed Kendall's tau between prenatal loss and familiar inbreeding coefficient and between neonatal loss and inbreeding coefficient and find a non-significant correlation. Instead, they find that the deleterious effects of inbreeding are quite staggering on survival to age 10, excluding prenatal and neonatal deaths (Alvarez *et al.* 2009).

9.3 Chapter 9 key concepts

- Why does correlation analysis not have a dependent and an independent variable?
- Why does correlation not mean causation?
- Explain the bivariate normal distribution.
- When should you use a non-parametric test for correlation?
- Explain what it means (for the two variables considered) if the correlation between them is positive, negative, or not different from 0.
- When is Kendall's tau appropriate instead of Spearman's?

9.4 Chapter 9 exercises

All data sets are available for download at www.cambridge.org/9780521147088.

1. A medical anthropologist wants to determine if there is a significant correlation between age and how well one scores on an exam which tests cognitive functions. The test, which has a total possible score of 100, was given to an elderly cohort whose ages range from 67 to 88. Compute the Pearson and Spearman correlation coefficient to determine if there is a significant correlation between age and test score. The data set is called "age and cognitive test."
2. Using the data set called "social status," compute the following:
   1. For the entire sample the Pearson and Spearman correlation between (a) Number of tooth losses and number of abscesses, (b) Number of tooth losses and number of burial goods, and (c) Number of abscesses and number of burial goods.
   2. For males and females separately compute Kendall's tau between (a) Number of tooth losses and number of abscesses, (b) Number of tooth losses and number of burial goods, and (c) Number of abscesses and number of burial goods. If you MUST do it by hand, do remember that getting the ranks with a computer package is quite helpful.
3. Using the data set called "seasonal workers," compute the following:
   1. The Pearson and Spearman correlation for BMI and both blood pressure measures.
   2. Kendall's tau for BMI and both blood pressure measures, by nationality of origin.

10 Simple linear regression

As most statistics textbooks will tell you, regression analysis has many similarities with correlation analysis: both statistical techniques deal with roughly the same type of data, and both require a visual inspection of the data's scatter plot to determine if the variables have the requisite distribution. This is all true; however, I would prefer to introduce regression analysis by reminding us of analysis of variance. In section 6.1 I wrote: "When we think of model I ANOVA in terms of dependent and independent variables it is easier to understand that ANOVA is **a form of regression analysis**, in which we seek to explain the dependent variable as follows: $\widehat{Y}_{ij} = \mu + \alpha_i + \varepsilon_{ij}$, where $\widehat{y}_{ij}$ = the predicted y for the jth observation which is in the ith group, μ = the mean of all observations in the study regardless of the group to which they belong, α_i is the treatment effect of being in group i^{th}, and ε_{ij} is the normal random variation found within the ith group which is measured in the jth observation." When we are doing an ANOVA the dependent variable Y is measured in subjects who belong to two or a few more groups. The key issue is that the independent variable is categorical (two genders, three age groups, four community centers, etc.). In simple linear regression we also measure a dependent variable (Y) and an independent variable (X) on the subjects, but the independent variable is not categorical but is rather continuous or discontinuous numeric with many levels. When we do ANOVA we attempt to understand and partition the variation of the Y in terms of the effect of the X on it. In simple linear regression analysis we attempt to do the same, with the caveat that the independent variable has more than a few values. For example, if we are investigating how income (dependent variable) is affected by education (independent variable), and education ranges from 0–24 years, we could approach this as an ANOVA problem. However, if we did, we would have in our hands 24 groups! Regression analysis allows us to investigate this issue in a much simpler manner.

This chapter covers the topic of simple linear regression; that is, regression analysis with only one independent variable (thus simple as opposed to multiple) and in which the relation between the X and the Y is linear (as opposed to curved). The independent or **predictor variable** is the one we use to understand, explain, and predict the behavior of the dependent variable. The researcher manipulates or controls the independent variable, in order to observe the response of the dependent one. Thus, the main distinction between the X and Y is that the former can be controlled, or at least is measured without error by the researcher. The latter is free to vary, and is simply recorded (not manipulated). According to Draper and Smith (1998), the distinction between **predictor** and **response** variables is dependent upon the research project's purpose. A variable which at some point is used as a response may at a later point be the predictor (Draper and Smith 1998).

The purposes of regression analysis can be summarized as follows: (1) Mathematical description of the relation of the X and the Y, where X is the independent variable and Y is the dependent variable; (2) Prediction of Y, based on knowledge of the relation between X and Y. Although we may predict the dependent variable given a value of the independent one, we do not mean that the X explicitly and physically "causes" the Y. In other words, a regression analysis will not explain the physical, biological, or cultural mechanisms linking the two variables ("the cause"). It will simply say that X can predict the behavior of Y. An example will clarify this: let us presume that an anthropologist is studying the material items in households in a rural community in order to predict the household's income. For the sake of argument, let us also assume that this is a multiple regression exercise, in which there is one dependent variable (income), and a number of independent variables (e.g. number of electrical appliances, number of dogs in the house, size of the house, etc.). By means of regression analysis, the anthropologist finds that the independent variable which best predicts household income is the number of dogs. This conclusion does not imply that the more dogs a household has, the larger its income will become. This conclusion does not say that "X causes Y." It simply says that the best predictor of the behavior of the Y out of a number of independent variables is that particular X.

Because regression analysis is a multi-step procedure instead of a one-hypothesis test, I describe the entire process first, and then take a step-by-step approach with an actual data set.

10.1 An overview of regression analysis

In regression analysis, sample data are used to test a hypothesis about the *parametric relation* between two variables. The sample statistic we use to test the hypothesis about the population is the slope, usually denoted by b. The population parameter about which we test a hypothesis is the population slope, usually denoted by β (beta). The slope measures the strength and direction of the effect of X on Y. Regression proceeds in the following fashion:

1. *The data are plotted and inspected to detect violations of the linearity and homoscedasticity assumptions*. Please remember that we are discussing linear regression for which a linear relation between the X and the Y is an assumption. By convention, the independent variable is plotted in the horizontal (**abscissa**) axis, and the dependent variable in the vertical (**ordinal**) axis. These axes are frequently referred to as the X and Y axis respectively. Thus, each observation will be marked with a dot at the intersection of its X and Y values. The entire sample will form a scatter of points. If this scatter has a shape which is not linear, that is, if the scatter forms a curve or a "fan," then linear regression cannot be applied to the data set. A fan shape would indicate that as the

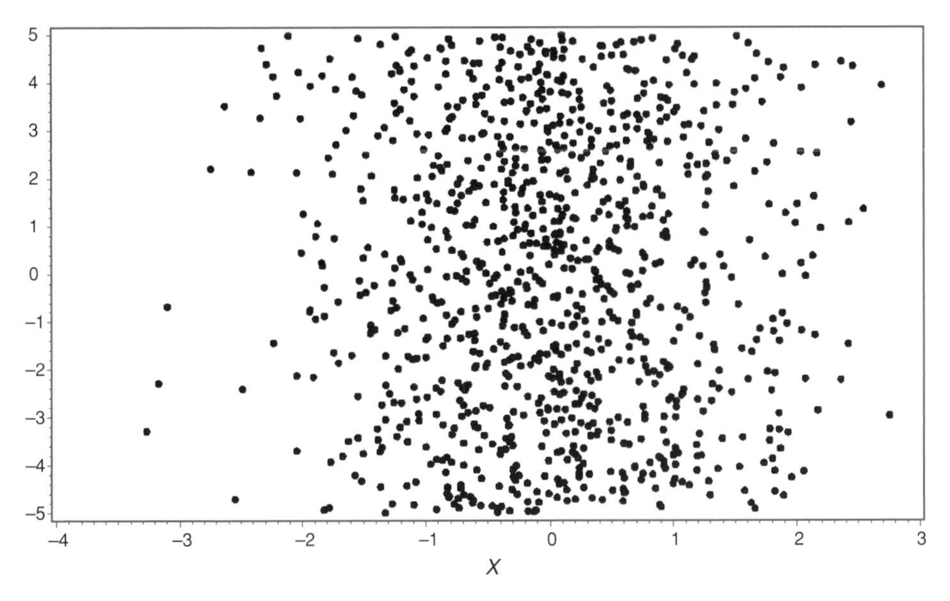

Figure 10.1 A scatter plot of two unrelated random variables.

values of the X increase, the variance of the Y increases, so that the homoscedasticity assumption is violated. Figure 10.1 shows the scatter plot of two randomly generated variables. The scatter, which shows no clear increasing or decreasing pattern, is what would be expected if two unrelated, random variables were plotted together. The scatter plot does not indicate that the assumptions of linearity or homoscedasticity are violated.

2. *The relation between the X and the Y is described mathematically with an equation.* The reader probably recalls from Middle School that a straight line is described by an equation which does not have terms raised to any power (such as X^2 or X^3). **Curvilinear regression** (not covered here), incorporates such terms because its regression equation describes a curved line. Now, if we are to compute a linear equation that best describes a scatter of points, which line should we describe? We can trace lines across any scatter of points, in any direction. *The line that we want is the line that is closest to all points at once*, and is the one obtained with the "**least-squares**" technique. The idea behind the least-squares technique in the computation of the regression equation is discussed in the section on residual analysis, below. We know something else about this line: it must go through the means of both variables. The formula for the computation of the sample equation is shown in formula 10.1.

Formula 10.1

Sample formula of the regression equation.

$$\hat{Y} = a + b(X) + \varepsilon$$

where: $\hat{Y}$ (Y hat) is the predicted value of Y, a is the point at which the line intercepts the Y axis and b is the slope of the line. We compute the equation's elements as follows:

$$b = \frac{SP_{XY}}{SS_X}$$

$$a = \bar{Y} - (b)(\bar{X})$$

Where the cross-products sum of squares is $SP_{XY} = \sum XY - \frac{(\sum X)(\sum Y)}{n}$ and the SS_x is shown in formula 9.1

The slope of the equation is going to be crucial for our understanding of the relation between the X and the Y. As opposed to the correlation coefficient, which can only take values from -1 to $+1$, the slope is measured in the units of the Y. The slope is going to tells us precisely the effect of a one-unit change in X on the Y.

3. *The regression analysis is expressed as an analysis of the variance of* Y, where the total "pie" of variation is the SS_Y. You will notice that much of the terminology we use in this chapter is the same as we used in the ANOVA chapter. In regression analysis we will partition the SS_Y into the SS explained by the model (where the model is the regression model) and the SS not explained by the model, which may be called SS error or SS unexplained. We will also compute mean squares (MS), obtained by dividing the SS by their df. Using the MS, we will compute an F ratio (by dividing the MS regression by the MS error) to test the null hypothesis that X does not explain a significant amount of Y's variation. We will also compute **the coefficient of determination or R^2** which expresses in a scale of 0 to 1 the proportion of the total variation of Y explained by X. If $R^2 = 0$, then X does not explain any of the variation of the Y, and if it is equal to one, then X explains all the variation of the Y. Notice that the R^2 does not test a hypothesis; the null hypothesis is tested with the F ratio.

4. *The null hypothesis that the parametric value of the slope is not statistically different from zero is tested.* The component of the sample equation which expresses the strength and direction of the relation between the X and Y is the slope (the Y intercept only shows where the line crosses the Y axis, but does not inform us about the statistical significance of the relation between the variables). If b is positive and statistically different from zero, then as X increases Y increases. If b is negative and statistically different from zero, then as X increases Y decreases. And if b is not statistically significantly different from zero, then the X does not affect and cannot be used as a predictor for the behavior of Y. Thus, we use the slope computed in our sample to test the hypothesis that the parametric slope is equal to zero. We test this hypothesis by computing a t score with the standard error of the slope (equation shown below) and compare it with the critical value of the t table at $df = n - 2$ (see Table 4.1).

For sake of completeness, the parametric regression equation (which uses Greek symbols) is shown in formula 10.2. Notice that the parametric equation does not have an error term in it.

Formula 10.2

The population regression equation.

$$Y = \alpha + \beta(X)$$

Where Y is known without error, hence the lack of the "hat" (ˆ) on top of it, and the lack of the error term, α is the parametric value at which the line described by the equation intersects the Y axis, β is the slope of the line. When we write down $\beta(X)$, we mean that the slope, multiplied by a particular X value, results in the known value of Y, in conjunction with the rest of the equation.

5. *The regression equation is used to predict values of Y.* This is done by substituting in the regression equation the X value whose Y we want to predict. The predicted values are usually given with confidence intervals, since the confidence associated with our predictions depends on the value of X entered in the equation. If the value of X is the mean of the independent variable or very close to it, our confidence is greatest and the confidence interval is smallest. If the value of X is very different from the mean, our confidence in the prediction is diminished, and the interval is wider.

6. *Lack of fit is assessed.* The analyst would like to determine if the regression model he has fit to the data is the best possible model that can be obtained. If a model explains 20% of the variance and another explains 50% of the variance of the dependent variable, the latter is more successful. A better model explains more aspects of the relation between the X and the Y than a model that explains fewer aspects of this relation. Let me explain this with an example. We know that if we consider the relation between human age and human growth, such relation for the second decade of life is roughly linear and positive: children who are 16 years of age tend to be taller than children who are 14, while children who are 13 tend to be taller than children who are 11. However, after age 20 humans do not continue to grow (with few exceptions, of course) but achieve a stable height. Much to their dismay, people after age 40 begin to lose height (according to the NIH, 1 cm is lost every 10 years after the age of 40; see http://www.nlm.nih.gov/medlineplus/ency/article/003998.htm). If we want to model with a regression equation the relation between age (X) and height (Y) in humans from age 10 until 100, we would need to consider that from age 11–20, as age increases, height increases; for this decade the slope of the equation would be positive. For the next 20 years as age increases there is no significant change in height, so the slope of the equation for this time should not be different from zero. After age 40 until death, as age increases height decreases; for this time period the slope of the equation should be negative. Thus, if we compute a regression equation that only incorporates a positive slope we would be producing a model that suffers from lack of fit because it leaves out the flat and negative aspects of the relation between X and Y. Such lack of fit might be obvious in the residual analysis: we might plot the residuals against the X and see a large and obvious curve. However, there is a simple test to determine if the model suffers from lack of fit, a test which can be used in data sets which have more than one observation

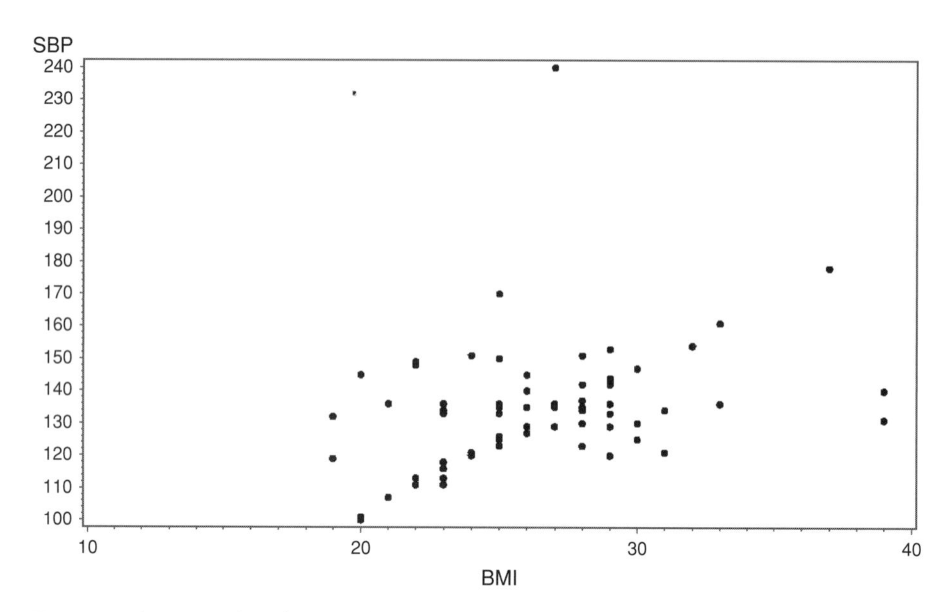

Figure 10.2. A scatter plot of BMI and SBP.

with the same X value. A lack of fit test will allow us to determine if all of the variation of the residuals is random (which should be the case if the model fits) or if there is a consistent non-random aspect to the residuals (the curved aspect of the relation between the X and the Y). Please note that if we improve our model after we have determined that the initial model suffers from lack of fit then our R^2 will likely improve because we have explained a greater proportion of the variation of the Y.

7. *The residuals are analyzed.* The question at this point is: how well does the model fit the data? If the fit is poor, what can be done to improve it? The two usual responses to the last question are: transform the data to make the relation between X and Y linear, or use non-linear regression.

10.2 Regression analysis step-by-step

We will now cover all these steps with the data set called "seasonal workers." These data include basic health measures and country of origin recorded in 75 migrant workers. At this point, let us ask if the person's body mass index (BMI) is a good predictor of systolic blood pressure (SBP). Thus, let us make BMI the X and SBP the Y. We follow the steps outlined before.

10.2.1 The data are plotted and inspected to detect violations of the linearity and homoscedasticity assumptions

Figure 10.2 shows the scatter plot of BMI (in the horizontal line) against systolic blood pressure (in the vertical line). The variables appear to be related in a linear manner. They also indicate that as BMI increases, so does SBP. There is also no evidence that as the X increases in value the variation of the Y increases (resulting in a "fan-like" shape).

10.2.2 The relation between the *X* and the *Y* is described mathematically with an equation

The formulae necessary for the computation of the regression equation are shown in formula 10.1. We need the following quantities. Keep in mind that X = BMI and Y = SBP.

$$n = 75, \quad \bar{X} = 26.36, \quad \bar{Y} = 134.5,$$
$$\sum X = 1977, \quad \sum X^2 = 53{,}395,$$
$$\sum Y = 10{,}090, \quad \sum Y^2 = 1{,}384{,}134$$
$$\sum X^*Y = 268{,}021$$
$$SP_{XY} = 268{,}021 - \frac{1977^*10{,}090}{75} = 2048.6$$
$$SS_x = 53{,}395 - \frac{1977^2}{75} = 1281.28$$
$$b = \frac{SP_{XY}}{SS_X} = b = \frac{2048.6}{1281.28} = 1.598 \approx 1.6$$
$$a = \bar{Y} - (b)(\bar{X}) = a = 134.5 - (1.6)(26.36)\ 92.324$$
$$\hat{Y} = 92.324 + 1.6\,(X) + \varepsilon$$

This equation tells us that for every unit increase in BMI (say from 25 to 26) we can expect to see an increase in systolic blood pressure of $\approx$ 1.6 units. We can also see that the equation line intercepts the Y axis at the value of 92.324.

10.2.3 The regression analysis is expressed as an analysis of the variance of *Y*

A complete regression analysis should include an ANOVA of the dependent variable. The purpose here is to partition the variation of the Y, and determine: (1) If X explains a significant amount of variation of the Y. This is indeed a null hypothesis (which is not written in parametric terms), and is tested with an F ratio. (2) What proportion of the total variation of the Y can be explained by the X.

We are therefore going to construct an analysis-of-variance table which will have two sources of variation: the regression model and error (that part of the variation of the Y which cannot be explained by the X). The table will show three sums of squares: SS total, SS regression, and SS error. By definition, these sums of squares are additive (the total pie of variation is sliced into two pieces, regression and error). The table will also have degrees of freedom associated with the SS regression and the SS error. As such, they will be referred to as regression and error df. We will compute the mean square due to regression (MS regression) by dividing the SS regression by its df, and the mean square due to error (MS error) by dividing the SS error by its df. We will compute an F

ratio by dividing the MS regression by the MS error, and compare it with the table of F ratios (see Table 6.1), using the associated df to find the critical value. If the test F ratio is greater than or equal the critical value at $\alpha = 0.05$, the null hypothesis that X does not explain a significant part of the variation of Y is rejected. Fortunately, the computations necessary for constructing this table were already obtained when the regression equation was computed. The regression analysis ANOVA table, with its formulae is shown below.

Formula 10.3

Formulae for computation of ANOVA in a simple regression analysis.

Source	df	SS	MS	F
Regression	1	$\frac{(SP_{XY})^2}{SS_X}$	$\frac{SSreg}{1}$	$\frac{MS\,reg}{MS\,error}$
Error	$n-2$	$SS_Y - SS\text{ reg}$	$\frac{SS\,error}{n-2}$	
Total	$n-1$	SS_Y		

We begin by computing the SS total, or the $SS_y = 1384{,}134 - \frac{10{,}090^2}{75} = 26{,}692.667 \approx 26{,}693$. The total degrees of freedom are $75 - 1 = 74$.

Next we compute the SS regression: $\frac{(SP_{XY})^2}{SS_X} = \frac{2048.6^2}{1281.28} = 3275.444$, with degrees of freedom = 1. The SS error is obtained by subtraction: SS error = 26,693–3275.444 = 23,417.556, with degrees of freedom = 75 – 2 = 73.

Next we compute the mean square error which is $MS\,error = \frac{23{,}417.556}{73} = 320.78$ and the MS regression, which is MS regression = $\frac{3275.444,}{1} = 3275.444$. Finally we compute the F ratio. Our null hypothesis is that BMI does not explain a significant part of the variation of systolic blood pressure. $F = \frac{MS\,reg}{MS\,error} = \frac{3275.444}{320.78} = 10.21$ with degrees of freedom 1 and 73. The critical values for df 1 and 70 are 3.98 for $\alpha = 0.05$ and 7.01 for $\alpha = 0.01$ (see Table 6.1). We reject the null hypothesis with a great degree of confidence: X does explain a statistically significant part of the variation of Y. But how much? To answer this question we turn to the **coefficient of determination or R^2** which expresses in a scale of 0 to 1 the proportion of the total variation of Y explained by X.

Formula 10.4

Formula for the coefficient of determination or R^2.

$$R^2 = \frac{SS\,regression}{SS\,total}$$

As you can see, the coefficient of determination tells us what proportion of the pie of variation of the Y is explained by the regression model. In our example the coefficient of determination is $R^2 = \frac{3275.444}{26{,}693} = 0.12$. Therefore, 12% of the variation of systolic blood pressure can be explained by variation in the subjects' BMI. We present all of our ANOVA calculations in a standard ANOVA table (Table 10.1).

10.2.4 The null hypothesis that the parametric value of the slope is not statistically different from 0 is tested

We need to compute a t score by dividing the slope by the standard error of the slope. We compare the slope with the critical values provided by the t table at $\alpha = 0.05$ with $df = n - 2$.

Formula 10.4

Formula for the t score for H0: $\beta = 0$.

$$t = \frac{b - 0}{S_b}, \quad \textit{where}$$

$$S_b = \sqrt{\frac{MS\ error}{SS_X}}$$

With degrees of freedom $= n - 2$.

In our example $S_b = \sqrt{\frac{320.78}{1281.28}} = 0.50$. Thus $t = \frac{1.6}{0.50} = 3.2$. The critical value at $df = 60$ at $\alpha = 0.05$ is 2. Therefore we reject the null hypothesis. If you do this example with SAS or PASW you will obtain a more precise p value. According to both computer programs the probability is <0.0021.

10.2.5 The regression equation is used to predict values of *Y*

When predicted values of Y are computed, the meaning of the value of the slope can be easily grasped: the slope is the amount by which Y will be changed, if the value of the X is increased by one unit. Although the complete equation is $\hat{Y} = 92.324 + 1.6(X) + \varepsilon$ the error term is not considered for computation purposes of the predicted value. The researcher simply acknowledges that there is error in the calculation, and accompanies the prediction with confidence intervals. Let us first compute predicted values of Y for the following values of X: 26, 27, 17, and 18. Thus, if:

X	$\hat{Y} = 92.324 + 1.6(X)$	then $\hat{Y}$
26	$\hat{Y} = 92.324 + 1.6(26)$	133.924
27	$\hat{Y} = 92.324 + 1.6(27)$	135.524
17	$\hat{Y} = 92.324 + 1.6(17)$	119.524
18	$\hat{Y} = 92.324 + 1.6(18)$	121.124

Please note that the difference between 135.524 and 133.924 and between 121.124 and 119.524 is 1.6, which is the value of the slope. Therefore, as we increase X by one (from 26 to 27), we increase $\hat{Y}$ by 1.6. When reporting prediction results, it is usual to accompany the prediction with the confidence interval for an individual $\hat{Y}$ for a specific X. The formula for this confidence interval is shown in formula 10.5.

Formula 10.5

Confidence interval (CI) for an estimated $\hat{Y}$ for a given value of X.

95% CI for an individual $= \hat{Y} \pm (S_{\hat{Y}})(t_{0.05,df})$, where

$$S_{\hat{Y}} = \sqrt{MS\ error * \left[1 + \frac{1}{n} + \frac{(X_i - \bar{X})^2}{SS_x}\right]}$$

where $S_{\hat{Y}}$ is the standard error for the individual $\hat{Y}$, X_i is the X for which we are predicting the $\hat{Y}$, and $t_{0.05,\text{df}}$ is the critical value in the t table at $\alpha = 0.05$ for df $= n - 2$.

We proceed to compute the standard error and confidence interval for the predicted values already computed. We know that the MS error $= 320.78$, $n = 75$, df $= 73$, $\bar{X} = 26.36$, $SS_X = 1281.28$, and that $t_{0.05,73} = 1.993$ (taken from the complete table of t scores).

We compute the confidence interval (CI) for an estimated $\hat{Y}$ for a given value of X as follows:

X	$\hat{Y}$	$S_{\hat{Y}} = \sqrt{MS\ error * \left[1 + \frac{1}{n} + \frac{(X_i - \bar{X})^2}{SS_X}\right]}$	$(S_{\hat{Y}})(t_{0.05,df})$	$\hat{Y} \pm (S_{\hat{Y}})(t_{0.05,\text{df}})$
26	133.924	$S_{\hat{Y}} = \sqrt{320.78 * \left[1 + \frac{1}{75} + \frac{(26-26.36)^2}{1281.28}\right]} = 18.03$	35.934	97.99–169.858
27	135.524	$S_{\hat{Y}} = \sqrt{320.78 * \left[1 + \frac{1}{75} + \frac{(27-26.36)^2}{1281.28}\right]} = 18.03$	35.934	99.59–171.458
17	119.524	$S_{\hat{Y}} = \sqrt{320.78 * \left[1 + \frac{1}{75} + \frac{(17-26.36)^2}{1281.28}\right]} = 18.63$	37.125	82.394–156.654
18	121.124	$S_{\hat{Y}} = \sqrt{320.78 * \left[1 + \frac{1}{75} + \frac{(18-26.36)^2}{1281.28}\right]} = 18.51$	36.89	84.234–158.014

Please notice that the standard error of values far removed from the mean (for example, $X = 17$ and $X = 18$) is larger than for values close to the mean (values of $X = 26$ and $X = 27$).

This section ends with a plot of the observed data with the equation line superimposed (Figure 10.3). There is a topic that on the surface is rather obvious, and is thus frequently left out in statistics textbooks: the line which is described by the equation consists of

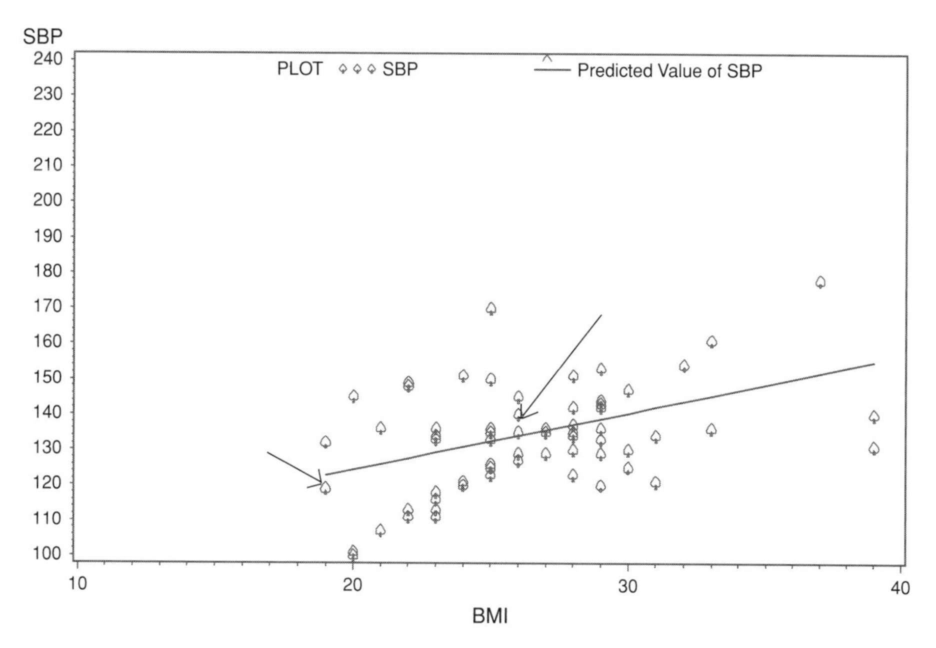

Figure 10.3 A scatter plot of BMI and SBP and the regression line. The arrows are explained on page 221.

the predicted values generated by the equation. For example, the fourth subject in our data set has a BMI of 26, an observed SBP of 140, and a predicted SBP of 133.924. The observed data point for this subject is noted in Figure 10.3. By simple inspection you can see that the observed point falls over the predicted value, and that the predicted value *is* the value for that *X* on the equation line.

10.2.6 Lack of fit is assessed

If you have repeated values of *X*, you should take the time to do a lack-of-fit test to determine if your model could be improved. If you have explained all of the variation of *Y due to X* then your residuals should have a totally random behavior when plotted against the *X*. However, if the relation between *X* and *Y* is more complex than is acknowledged in your current regression model, when you plot the residuals against the *X* you might see a systematic behavior in their scatter (for example, a curve). The lack-of-fit test is going to tell you if the residuals have such a systematic departure from randomness or if they are absolute random observations, in other words if they are pure error. To do a lack-of-fit test we partition the SS error into two parts: lack-of-fit SS and pure-error SS. In a nutshell, the SS pure error is going to quantify the amount of normal and random variation in the *Y* in different observations with the same value of *X*. The lack-of-fit SS is obtained by subtracting from the SS error the SS pure error. With these sums of squares we compute mean squares and test the null hypothesis that there is no lack of fit (or that all variation in the residuals is pure error). The necessary formulae are shown in formulae 10.6.

Formulae 10.6

Formulae necessary for the computation of pure error tests.

SS pure error = SSPE $= \sum_1^m SS_{Yi}$ (where SS_{Y_i} is the sum of squares for each X – from 1 to m – which has repeated values of Y. If there are only 2 values of Y for an X, the SS for each group of repeated values of X is: $SS_Y = \frac{(Y_1 - Y_2)^2}{2}$. The total SSPE is the sum of the individual SS_{Y_i}. The degrees of freedom are $\sum n - m$ where n refers to the number of replicates per X.

SS lack of fit = SS error − SS pure error.

MS pure error $= \dfrac{SS\ pure\ error}{\text{df}}$

MS lack of fit $= \dfrac{SS\ lack\ of\ fit}{\text{df}}$

F lack-of-fit test $= \dfrac{MS\ lack\ of\ fit}{MS\ pure\ error}$ with the corresponding degrees of freedom.

In our example we have several X values for which we have repeated measures of Y. For example, for BMI 19 we have two values of Y (132 and 119), for BMI 20 we have three values of Y (145, 100, and 101), and so forth. For these two values of X the SS_Y is 84.5 ($SS_y = \frac{(132-119)^2}{2}$) and 1320.666 ($41{,}226 - \frac{346^2}{3}$) respectively. The sums of squares and degrees of freedom for each value of X are:

BMI	SS_Y	df
19	84.5	1
20	1320.66	2
21	420.5	1
22	1616	4
23	708	6
24	771.33	5
25	2053.66	11
26	224.8	4
27	8562	3
28	472	6
29	798.22	8
30	392.75	3
31	84.5	1
32	0	1
33	312.5	1
39	40.5	1

Therefore the SS pure error is 84.5 + 1320.66 + ... + 312.5 + 40.5 = 17,862. The degrees of freedom are 1 + 2 ... + 1 + 1 = 58. Thus the SS lack of fit is 23,417.556 – 17,862 = 5555.556 with df = 73 – 58 = 15. The mean squares are: MS lack of fit:

Table 10.1 ANOVA table of regression analysis (X = BMI, Y = SBP).

Source	df	Sum of squares	Mean square	F value	p
Regression	1	3275.444	3275.444	10.21	0.0021
Error	73	23,417.556	320.78		
Lack of fit	15	5555.556	370.37	1.20	0.2962
Pure error	58	17,862	307.96		
Total	74	26,693			
$R^2 = 0.12$.					
Variable	df	Parameter estimate	Standard error	t value	p
Intercept		92.324			
BMI	1	1.6	0.50	3.20	0.0021

$\frac{5555.556}{15} = 370.37$ and MS pure error: $\frac{17{,}862}{58} = 307.96$. The F ratio for testing the null hypothesis that there is no lack of fit is: $F = \frac{370.37}{307.96} = 1.20$, which is not significant for these degrees of freedom (see complete ANOVA table in Table 10.1). Therefore, there is no significant lack of fit. I know you are as relieved as I am.

10.2.7 The residuals are analyzed

After the regression equation has been computed, we have for each observation the observed X and Y values, and the $\hat{Y}$. If the predicted value is subtracted from Y, then we obtain ε, the error in our prediction. This is the portion of Y's variation that remains unaccounted for in regression analysis. Figure 10.3 shows two data points, one above and one below the regression line. The difference between the observed Y and the expected $\hat{Y}$ is positive for the point which is above the line, while the difference is negative for the point below the line. Indeed, half of the data points are below and half of the data points are above the equation line so that if we were to add all the residuals the total would be zero. This problem is easily solved if we square the differences before we add them. *The equation line obtained by the* ***least-squares method*** *is that which minimizes the sum of squared residuals.*

It is of extreme importance that the residuals be plotted against the X because they may reveal if the regression model is not appropriate. If the model fits, the scatter of the residuals should present no distinct shape. If, however, the residuals have a curved shape, or if the residuals look like a fan, then the data may have to be transformed or a non-linear model may have to be applied. Please note that even if a model is significant, it could still suffer from lack of fit. For example, if the relation between X and Y is curvilinear, and a linear regression is applied to the data, the model could very well be significant because it explains the portion of X and Y's relation which is linear. Continuing with our example, the residuals of the regression analysis of SBP on BMI are shown in Figure 10.4. Although no fanning effect is obvious, there might be a suggestion of a curved relation.

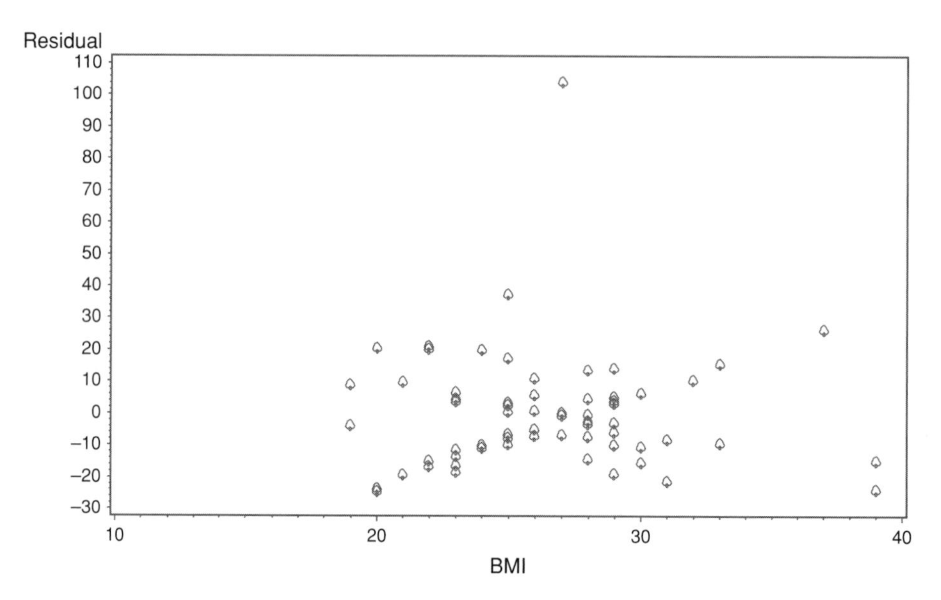

Figure 10.4 A scatter plot of the residuals against the X.

Besides a residual plot inspection, there are several other possible residual analyses which are available to the researcher to determine if the model could be improved. The researcher could:

1. Test the normality of the residuals with one of the tests covered in chapter five.
2. Plot the residuals to see if they follow a normal distribution.
3. Use any of the diagnostic tools offered by computer programs such as PASW and SAS which diagnose large residuals.

In our example above, two normality tests of the residuals showed that the residuals do not follow a normal distribution (Shapiro–Wilks $p < 0.0001$ and Kolmogorov–Smirnov $p < 0.0100$). Figure 10.5 shows a bar graph of the residuals, and indicates that the residuals are heavily skewed to the right. I think we can safely conclude that our residuals indicate that the model needs to be improved.

Practice problem 10.1

Let us regress diastolic blood pressure (DBP) on BMI following the steps outlined in the previous pages.

1. *The data are plotted and inspected to detect violations of the linearity assumption.* Figure 10.6 shows the scatter plot of BMI against diastolic blood pressure. The variables appear to be related in a linear manner. The plot indicates that as BMI increases, so does SBP.

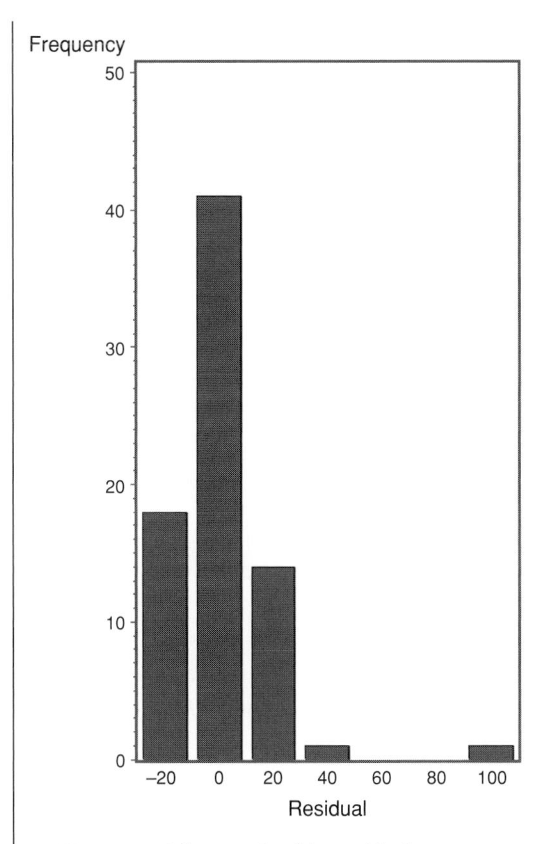

Figure 10.5 A bar graph of the residuals.

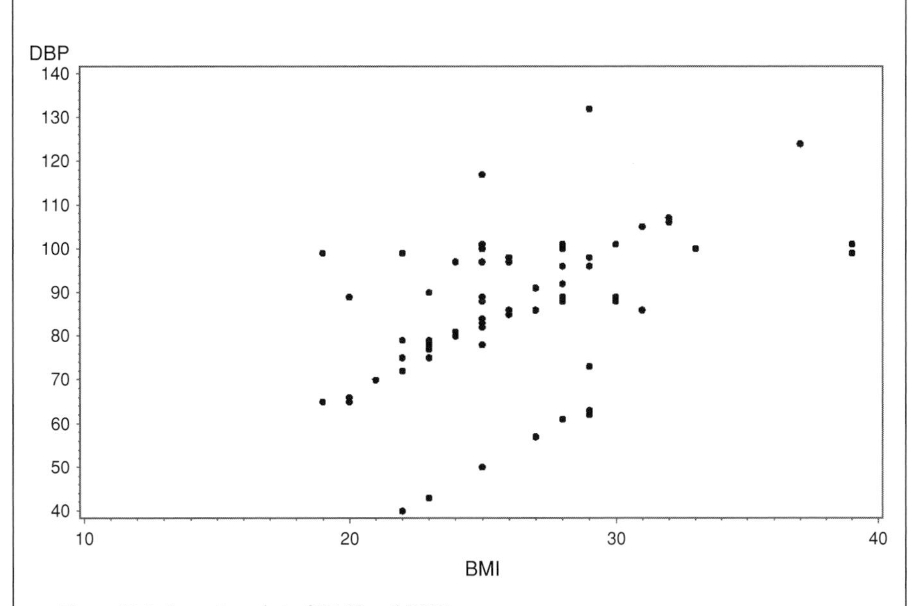

Figure 10.6 A scatter plot of BMI and DBP.

2. *The relation between the X and the Y is described mathematically with an equation.* We need the following quantities. Keep in mind that X = BMI and Y = DBP.

$$n = 75, \quad \bar{X} = 26.36, \quad \bar{Y} = 85.75,$$
$$\sum X = 1977, \quad \sum X^2 = 53{,}395,$$
$$\sum Y = 6431, \quad \sum Y^2 = 572{,}659$$
$$\sum X^* Y = 171{,}895$$
$$SP_{XY} = 171{,}895 - \frac{1977^* 6431}{75} = 2373.84$$
$$SP_x = 53{,}395 - \frac{1977^2}{75} = 1281.28$$
$$b = \frac{SS_{XY}}{SS_X} = b = \frac{2373.84}{1281.28} = 1.85$$
$$a = \bar{Y} - (b)(\bar{X}) = a = 85.75 - (1.85)(26.36) = 36.9$$
$$\hat{Y} = 36.9 + 1.85\,(X) + \varepsilon$$

The equation tells us that the equation line will cross the Y axis at 36.9 and that for every unit increase in X (BMI) there will be an increase of 1.85 in Y (DBP).

3. *The regression analysis is expressed as an analysis of variance of Y.*

We begin by computing the SS total, or $SS_Y = 57{,}2659 - \frac{6431^2}{75} = 21{,}222$. The total degrees of freedom are 75 – 1 = 74.

Next we compute the SS regression: $\frac{(SP_{XY})^2}{SS_X} = \frac{2373.84^2}{1281.28} = 4398.04$, with degrees of freedom = 1. The SS error is obtained by subtraction: SS error = 21,222 – 4398.04 = 16,823.96, with degrees of freedom = 75 – 2 = 73.

Next we compute the mean square error which is $MS\ error = \frac{16{,}823.96}{73} = 230.46$ and the MS regression, which is $MS\ regression = \frac{4398.04}{1} = 4398.04$. Finally we compute the F ratio. Our null hypothesis is that BMI does not explain a significant part of the variation of diastolic blood pressure. $F = \frac{MS\ reg}{MS\ error} = \frac{4398.04}{230.46} = 19.08$ with degrees of freedom 1 and 73. The critical values at 1 and 70 degrees of freedom are 3.98 for $\alpha = 0.05$ and 7.01 for $\alpha = 0.01$. We reject the null hypothesis: X does explain a statistically significant part of the variation of Y. The coefficient of determination is $R^2 = \frac{4398.04}{21{,}222} = 0.207$. Therefore, about 21% of the variation of systolic blood pressure can be explained by variation in the subjects' BMI. We present all of our ANOVA calculations in a standard ANOVA table. The table includes the lack-of-fit test which is discussed below.

Source	df	SS	MS	*F*	*p*
Regression	1	4398.04	4398.04	19.08	<0.0001
Error	73	16,823.96	230.46		
Lack of fit	15	2559.96	170.66	0.69	0.7797
Pure error	58	14,264	245.93		
Total	74	21,222			

4. *The null hypothesis that the parametric value of the slope is not statistically different from 0 is tested.* In our example $S_b = \sqrt{\frac{230.46}{1281.28}} = 0.42$. Thus our t score is: $t = \frac{1.85}{0.42} = 4.36$. The critical value at df $= 60$ at $\alpha = 0.05$ is 2. Therefore we reject the null hypothesis. If you do this example with SAS or PASW you will obtain a more precise p value. According to both computer programs the probability is <0.0001.
5. *The regression equation is used to predict values of Y.* We obtain predicted values and CI for the first two observations of our data set and for one very far away from the mean, let us say BMI $= 46$. We already determined that $= t_{0.05,\text{df}} = 1.993$. Please note that the confidence interval is much wider for the observation far away from the mean, as should be expected. Please note that the difference between the expected values of Y for two consecutive values of X is the value of the slope.

X	$\hat{Y}$	$S_{\hat{Y}} = \sqrt{MS\ error * \left[1 + \frac{1}{n} + \frac{(X_i - \bar{X})^2}{SS_X}\right]}$	$(S_{\hat{Y}})(t_{0.05,\text{df}})$	$\hat{Y} \pm (S_{\hat{Y}})(t_{0.05,\text{df}})$
29	90.55	$S_{\hat{Y}} = \sqrt{230.46 * \left[1 + \frac{1}{75} + \frac{(29-26.36)^2}{1281.28}\right]} = 15.322$	30.537	60.013–121.087
28	88.7	$S_{\hat{Y}} = \sqrt{230.46 * \left[1 + \frac{1}{75} + \frac{(28-26.36)^2}{1281.28}\right]} = 15.298$	30.489	58.211–119.189
46	122	$S_{\hat{Y}} = \sqrt{230.46 * \left[1 + \frac{1}{75} + \frac{(46-26.36)^2}{1281.28}\right]} = 17.404$	34.686	87.314–156.686

6. *Test for presence of lack of fit.* The lack-of-fit test yields non-significant results (shown in the ANOVA table). There is no evidence of lack of fit.
7. *The residuals are analyzed.* Both Shapiro–Wilks and Kolmogorov–Smirnov tests of normality rejected the null hypothesis of normality of the residuals. A scatter plot and a histogram of the residuals suggested that too many of the residuals were "clumped" around the mean of zero (probably because the data set was made up). Thus, we should look for ways to improve the model, whether by transforming the data or applying non-linear regression (Figures 10.7–10.8).

10.3 Transformations in regression analysis

As you know by now, you should look at plots at two points in time:

1. At first, before you do a regression, as a linear relation and a homoscedastic relation between the X and the Y are assumptions of linear regression.
2. After you have chosen your model, you should look at the residuals plotted against the X, and see if the behavior of the residuals tells you that you should try a transformation. If the residuals have a purely random behavior, then you have explained the variation of the Y which can be explained by the X in a *linear manner*. However, if the residuals plot in a non-linear manner against the X, then you could still explain more variation of the Y, but you need to quantify the relation between the variables in a non-linear manner. In other words, the X affects the Y in a non-linear and a linear manner. When you fit a linear regression, you have accounted for the latter but not the former.

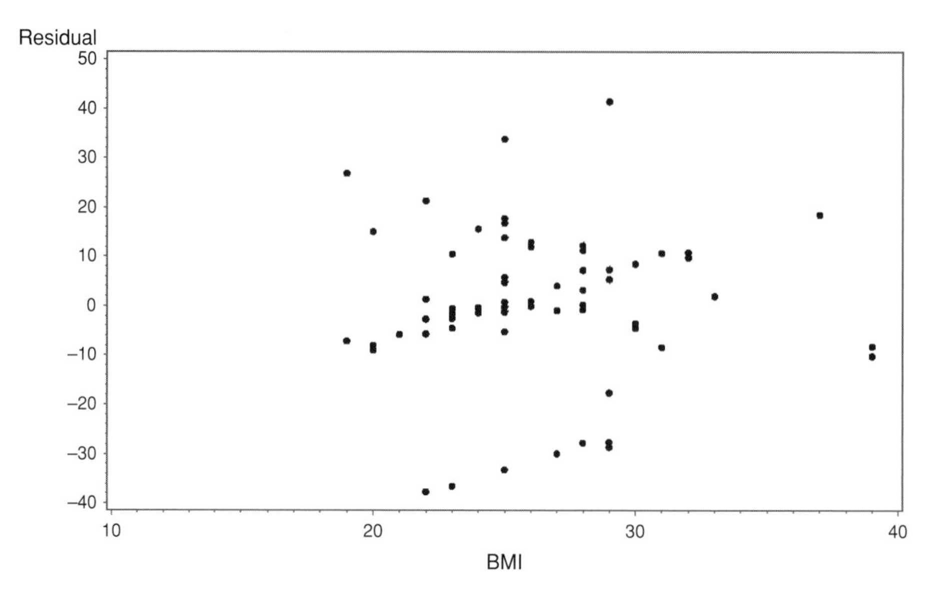

Figure 10.7 A scatter plot of the residuals.

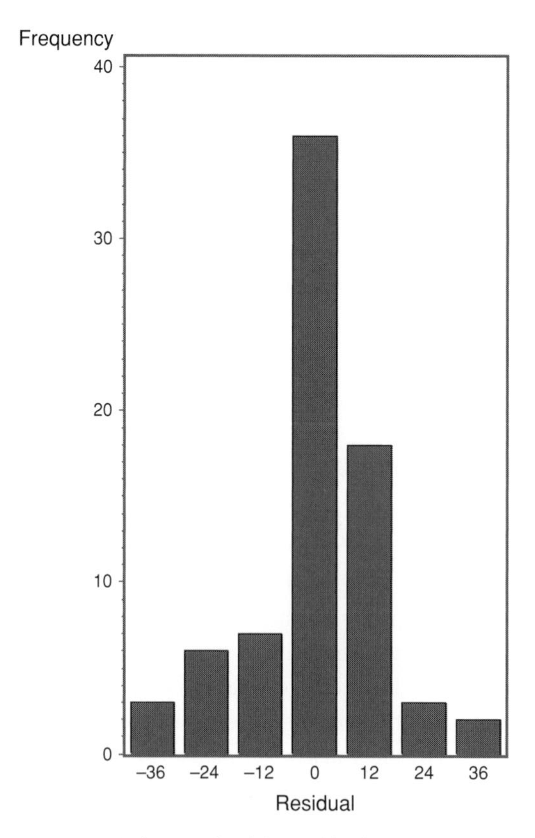

Figure 10.8 A bar graph of the residuals.

Therefore, you might be in a situation in which the residuals (and the lack-of-fit test) indicate that a better model may be needed. Or you might not even get to do the initial regression analysis because the raw data so blatantly violate the assumption of linearity. In either case you should try **data transformations**. Students may feel uncomfortable about transforming the data because they might feel that they are "fudging or faking" the data. However a data transformation is not going to allow you to "prove your hypothesis." Moreover, there is really no scientific necessity to employ the common linear or arithmetic scale to which we are accustomed. The decimal system of numbers is not the only "natural" one. Rather in transforming either *X* or *Y*, we aim at achieving a normal and homoscedastic distribution of points around the regression line. If a transformation straightens the relation between the *X* and the *Y*, and the systematic deviation of points around the line is reduced, then the transformation is worthwhile.

It is better to transform the *X* variable first before you attempt to transform the *Y*. After all, the dependent variable is one that is not controlled or manipulated by the researcher. Moreover if you transform the *Y*, you will have to consider that the predicted values are not for the *Y* but for the transformed *Y*. Finally, if you transform the *Y*, your R^2 will tell you the proportion of the variation of the transformed *Y* which is explained by the *X*, *not* the proportion of the variation of the *Y* explained by the *X*. It will be difficult to compare these two values.

If you use a logarithmic transformation and your data include values of 0, a technical problem arises, since the log of 0 is not defined. For that reason, you should first transform the variable by adding a 1 to it $(X + 1)$.

This is how you would write an equation in which you transformed the *Y* by using the log transform: $\widehat{logY} = a + bX + \varepsilon$. Other transformations might look like $\hat{Y} = a + b\sqrt{X} + \varepsilon$, $\hat{Y} = a + b_1X + b_2X^2 + \varepsilon$, $\hat{Y} = a + b\,(logX) + \varepsilon$, $\hat{Y} = a + b(\frac{1}{X}) + \varepsilon$, etc. It is often the case that you require an X^2 and an X^3. In such a situation you might want to add yet another term such as $X^2{*}X^3$. This term is called an **interaction term**. For each transformation, you should evaluate the model, the R^2, the lack of fit, and the residuals. In my experience the best way to find a satisfactory model that includes transformations is to ask PASW or SAS to do several transformations, re-do the regression analysis (including a residual analysis) and choose the best model based on the R^2 (assuming that the *Y* is not transformed), on a minimized MS error, and on the behavior of the residuals. However, several regression books provide guidelines on what kind of transformations might be best for your data. I constructed Figures 10.9–10.13 based on these sources (Dielman 2005; Draper and Smith 1998; Mendenhall and Sincich 2003; Tukey 1977). The figures show idealized violations to the assumption of linearity and homoscedasticity between the *X* and the *Y*. You might see such behavior of points either in the raw data or in the residuals of the initial regression analysis. Trust me: no real data look like data shown in textbooks, so it will not be as easy as these graphs suggest.

I will walk you through an exercise on transforming a data set but will not reproduce here all of the regression steps or plots. The data set is called "transformation" and it was made up with the aim of making it difficult to find an acceptable transformation. The plot of the raw data is shown in Figure 10.14, which clearly indicates that the assumption

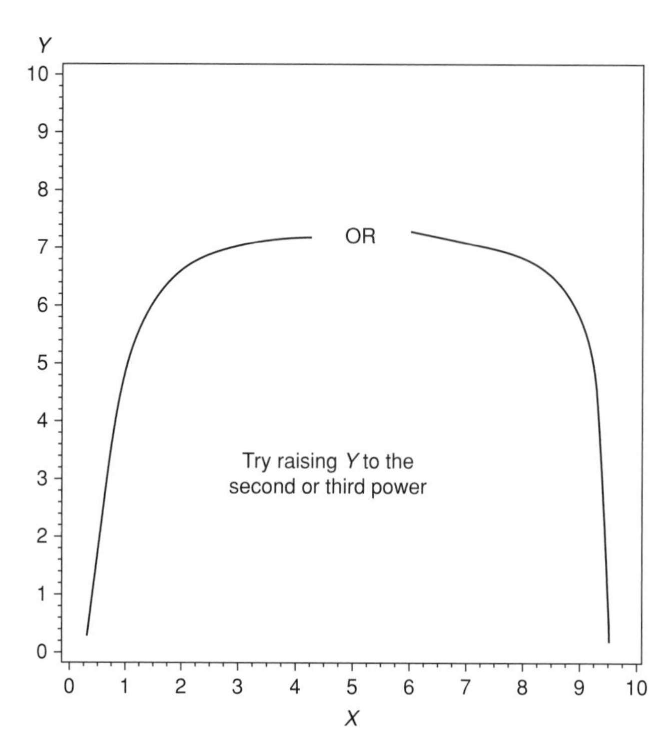

Figure 10.9 Violations to linear regression assumptions and suggestions on how to solve them: Y^2, Y^3, etc.

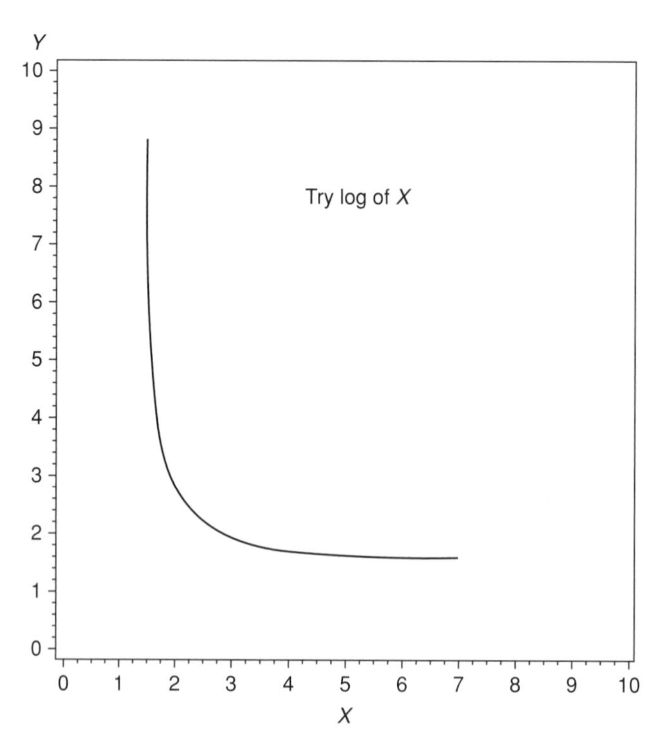

Figure 10.10 Violations to linear regression assumptions and suggestions on how to solve them: $\log X$.

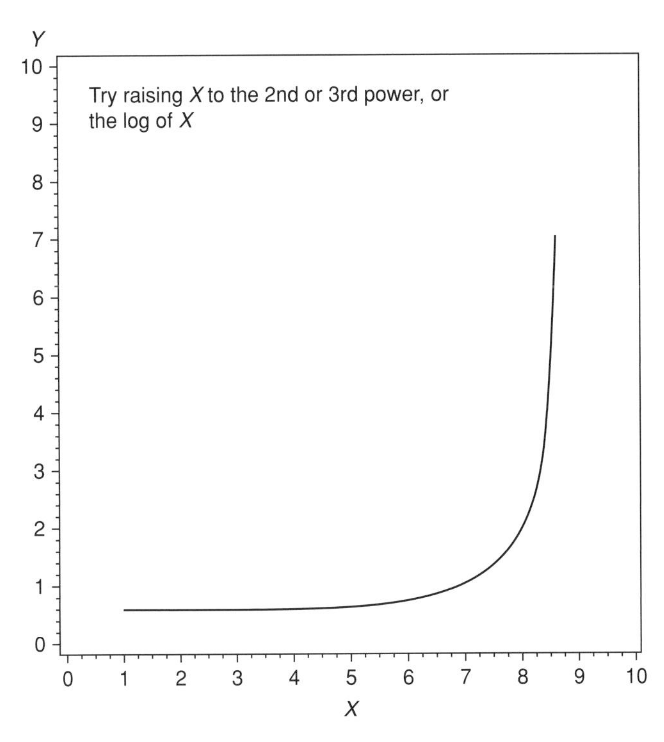

Figure 10.11 Violations to linear regression assumptions and suggestions on how to solve them: X^2, X^3, or $\log X$.

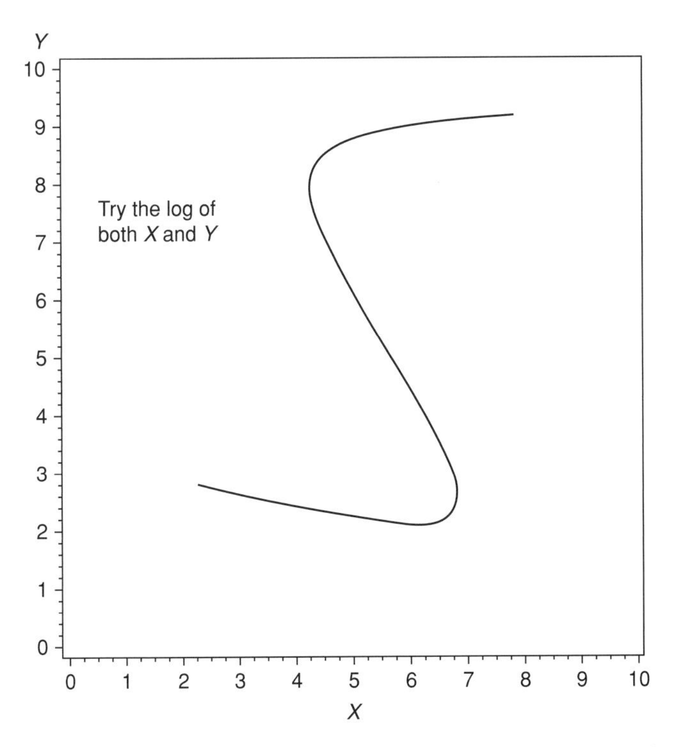

Figure 10.12 Violations to linear regression assumptions and suggestions on how to solve them: $\log Y$ or $\log X$ or both.

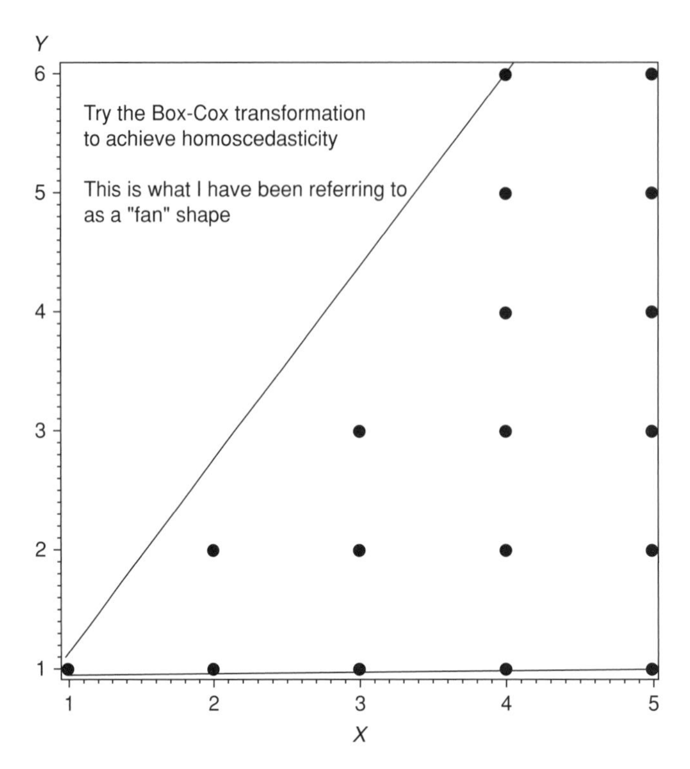

Figure 10.13 Violations to linear regression assumptions and suggestions on how to solve them: Box–Cox transformation.

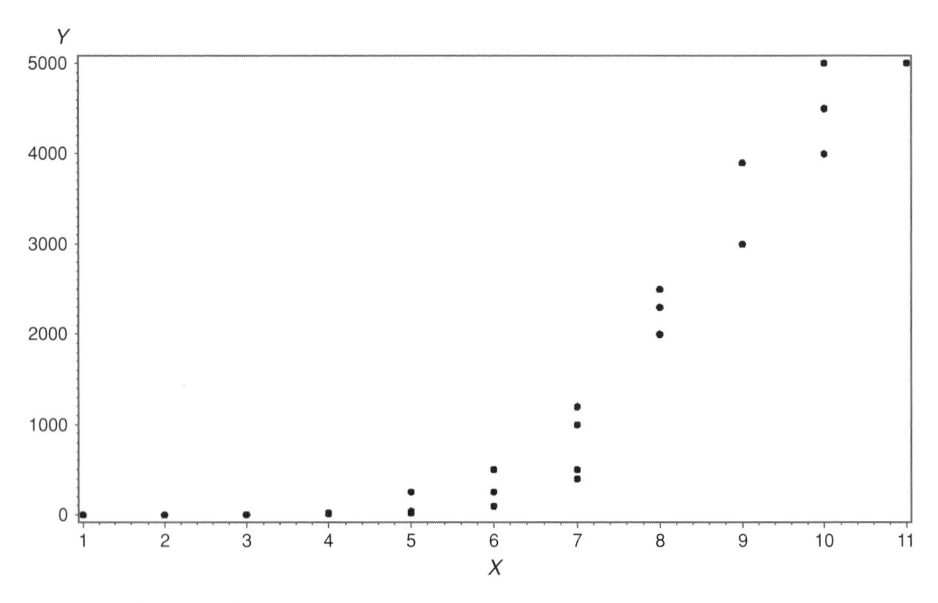

Figure 10.14 Scatter plot of raw X and Y for transformation exercise.

of a linear relation is violated in the data set. The regression analysis of the raw data gives us an R^2 of 0.734%, while the ANOVA and the slope are significant ($F = 77.324$, $df = 1, 28, p = 0.000$; $t = -4.2, p = 0.00$ respectively). This is an excellent example of a situation in which a model is significant while being inadequate because the data are in violation of the assumption of linearity.

Several transformations were tried and most of them produced plots that were just as bad as that of the raw data. Just to give you an idea of how the R^2 varies according to the transformation, here is a list of some:

1. Raw Y and natural log of X: $R^2 = 0.4$. The plot is very exponential-looking.
2. Raw Y and log base 10 of X: $R^2 = 0.4$. The plot is very exponential-looking.
3. Raw Y and X^2: $R^2 = 0.8$. Plot looks better though still not straight.
4. Raw Y and X^3: $R^2 = 0.9$. Plot looks almost straight.
5. Raw Y and $\sqrt{X}$: $R^2 = 0.5$. Plot looks better than the plots of the first two transformations, but not as good as those of the 3rd and 4th transformations.
6. Raw Y and $\frac{1}{X}$: $R^2 = 0.2$. Plot is quite ugly.
7. Log base 10 of Y against raw X: $R^2 = 0.8$. This is an interesting plot. It is better than #1, 2, 3, 5, and 6 but not as good as #4.
8. Natural log of Y versus X: $R^2 = 0.9$. This is an interesting plot. It is better than #1, 2, 3, 5, 6, and 7 but not as good as #4.

Therefore, we have two competing models:

1. $\hat{Y} = -293.096 + 4.509(X^3) + \varepsilon$
2. $\widehat{logY} = -0.843 + 0.989(X) + \varepsilon$

The ANOVA of both models is significant. The R^2 of the first model says that 90% of the variation of Y can be explained by X^3 while the R^2 of the second one says that 90% of the variation of the natural log of Y can be explained by the X. This is not the same thing. A residual analysis of both models led me to slightly favor the second model, but I still prefer to deal with the Y and not the transformed Y, so I favor the first model. You might want to try a model that has the X^3, the X^2, $X^3{*}X^2$, and $X^3{*}X$. Challenge yourself: can you find a better model? Please note that your final decision should consider everything, not only the R^2. If you get a modest increase in the R^2 with a very complex model, is that really worthwhile? Here you should consider what the residuals look like: if the residuals are pure error and lack any pattern then I would say that the complex model (with a minor increase in R^2) is worthwhile. But if the residuals of both models are basically equal, and if the R^2 increase is modest, I prefer the simple model. Finally, does the model *make sense?* A model with an impressive R^2 which nobody can interpret is not worthwhile. Please do not forget that you are doing regression analysis to answer an anthropological question, not to play in the computer lab with your geeky friends.

Simple linear regression in anthropology

What is the relationship between men's age and their levels of testosterone? Is this relation the same across different populations? Ellison *et al.* (2002) investigated the relation between males' age and their testosterone levels in several populations: USA, Paraguay, Congo, and Nepal. When all men were analyzed together, the authors obtained a highly significant equation which predicted that as age increases testosterone levels decrease. However, when the men were analyzed by population the regression equation was significant only for the USA and Congo. The authors showed that it is important to document cross-cultural variation in biological phenomena and discussed possible health applications of their results (Ellison *et al.* 2002).

10.4 Chapter 10 key concepts

- What does simple linear regression seek to explain?
- Explain the method of least squares.
- Parametric relation.
- X and Y axis.
- Expected shape of scatter plot in linear regression between two random variables.
- Linear relation.
- Coefficient of determination.
- Regression equation.
- Residual analysis (fanning, curved behavior).
- Transformations.
- How should you choose the best model with transformations?

10.5 Computer resources

- Both PASW and SAS have ample resources for regression analysis. Do explore the residual analysis options offered by both.
- Beware of the gratuitous use of a regression line in a scatter plot, which is the default in some programs. You should show the regression line only when it is relevant.

10.6 Chapter 10 exercises

All data sets are available for download at www.cambridge.org/9780521147088.

1. Use the data set analyzed in this chapter, and regress DBP on age.
2. Use the data set analyzed in this chapter, and regress SBP on age.

3. Use the data set called "women of two religions" and do the following regression analyses. Do it for the entire sample and for both subsamples separately.
   a. Children produced regressed on height.
   b. Children produced regressed on weight.
4. Use the data set called "menarche" and regress number of children produced on:
   a. Age at menarche.
   b. Age at first pregnancy.
   c. Number of pregnancies.

11 Advanced topics in regression analysis

This chapter covers topics which are frequently found in another category of textbooks directed to a different audience than that for which this textbook is written. Indeed, there are numerous and excellent "second-level" textbooks on advanced regression analysis (Dielman 2005; Draper and Smith 1998; Hocking 2003; Mendenhall and Sincich 2003; Ryan 2009); my purpose is not to compete with them in this single chapter. My purpose is limited to introducing the reader to two advanced regression applications frequently used in anthropology: multiple and logistic regression analysis.

11.1 The multiple regression model

The transition from simple to multiple regression analysis is not terribly difficult but it is certainly fun. In a multiple regression model we have one dependent variable and several independent variables which are *potential* predictors of the former. The art of regression model building resides in the analyst's particular choice of independent variables and their transformations into a model that is satisfactory. I call it an art because there is much room for individual choice. It is always fun to ask students to explain why they chose a model, and to hear different proposals for a better model. Not always is there an obvious best choice; sometimes there is considerable room for disagreement.

The best model is one that makes sense to the analyst and which explains as much variation of the Y with as few as possible Xs. The reason we want few variables is that if we are only guided by a desire to increase the R^2 and thoughtlessly add variables into the model, then the equation will be impossible to interpret. In addition, if a regression model has practical implications for policy planning, you do not want to push for a policy change based on complex and obscure models that nobody can interpret (even with a high R^2!). Instead, you want to advocate policy change based on simple and understandable models. There is another problem associated with adding too many variables to a model: the predictive ability of the model actually decreases. To guard unwise analysts against over-specifying a regression model we use the **adjusted R^2** (R^2 adj.) instead of the R^2 in multiple regression analysis. This statistic tells us the proportion of variation of the Y explained by the model, but it penalizes the analyst if she adds too many variables into the model. I have created models for teaching purposes with R^2 hovering around 100% but with R^2 adj. close to 0. Therefore from now on we will be using the more conservative R^2 adjusted rather than the plain R^2.

We will continue to use a least-squares approach to regression which still assumes that the Y and each of the Xs have a linear relation. Therefore, this assumption should be tested, as discussed in chapter 10. The multiple regression model, written in parametric and statistical notation is shown in formula 11.1.

Formula 11.1

Formula for the population and sample multiple regression equations.

The parametric equation is: $Y = \alpha + \beta_1 X_1 + \ldots \beta_k X_k$, and the statistical equation is: $\hat{Y} = a + b_1 X_1 + \ldots b_k X_k + \varepsilon$,
where each individual slope represents the unit change in Y per unit change in X, taking into account the effect of the other Xs.

In a multiple regression equation we remove the shared variance between X_1 and X_2, so that we can estimate the variance in Y caused by the *independent* effects of X_1 and X_2. If we were not to account for this joint variation, then we would overestimate the independent effect of each X on Y.

11.1.1 The problem of multicollinearity/collinearity

In multiple regression analysis we may face the problem that if two or more of the independent variables are highly correlated, then it becomes difficult to estimate reliably their slopes and the slopes' standard errors. Indeed, it is possible that we might (in error) fail to reject the null hypothesis that one of the slopes is equal to 0 when in fact it should be rejected, because the effect of that particular X is masked by the effect of the other one with which the former is highly correlated. There are three things we can do to avoid the problem of correlated independent variables, better known as the problem of **multicollinearity** or **collinearity**:

1. Do a **correlation matrix analysis** of all the variables. This is such a worthwhile endeavor, which will benefit your work on several levels that you should always try to do it. When I do a correlation matrix analysis I am reminded of what Gwilym Jenkins said about the process of model building: an analyst should fall in love with her data (not with the model) (Jenkins 1979). A correlation matrix analysis will give you the opportunity to study the behavior of all the variables at the same time. By examining a correlation matrix you might be able to see patterns in your data that you had not seen before. You might see, for example, that X_1 and X_2 are highly positively correlated with Y, while X_3 is highly negatively correlated with X_1 and X_2 but not with Y. What does this mean? An analysis of the correlation matrix will give you time to reflect on your data and to get to know them. For the purposes of model building specifically, you should be looking for the following:
   a. Which Xs are highly correlated with the Y? These Xs are precisely the variables which are likely to be good predictors of the Y. If you find that you only have low

correlations between the Y and the Xs do not expect to produce a model with a high R^2 adjusted.

b. Which Xs are highly correlated among them? If the independent variables are highly correlated you will not be able to include them all in the model. You will have to discard one out of two. Which one, you ask? Here you would consider which one is more highly correlated with the Y, and, of course, which one produces the better model. I explain how to choose the better model below.

2. Both PASW and SAS produce a measure of multicollinearity called **tolerance (TOL)**. Tolerance is computed by obtaining the R^2 without one of the Xs, such that tolerance is $1 - R^2$. If two variables are perfectly correlated, then tolerance $= 0$, Therefore, you should be wary of low values of tolerance, but what a low value is seems to change from statistics source to source. I have seen anything from 0.20 to 0.01 as the guideline for deciding what a low tolerance is.
3. Both PASW and SAS also compute a statistic called the **variance inflation factor (VIF)**, computed as VIF$= \frac{1}{TOL}$. You should be wary of high values of VIF, where all sources agree that a VIF higher than 10 is not good, while some sources indicate that a VIF of 5 or 4 is problematic.

11.1.2 The algebraic computation of the multiple regression equation

Most textbooks explain the computation of the slopes and intercept of a multiple regression equation using **matrix algebra**, which I did not assume readers knew when I wrote this book. Therefore, I think it would be unproductive to explain how the equations are computed in this manner. Rather, I will show how the equation is computed with the same level of algebraic complexity we have used thus far in the book. I do ask for your patience because it is not a quick and easy computation. If we have two independent variables (X_1 and X_2), the computation of the multiple regression equation proceeds as follows:

1. We regress X_1 on X_2. With this regression equation we predict $\hat{X}_1$ for each value of X_2. We obtain the errors in the prediction (differences between the observed X_1 and the $\hat{X}_1$) and multiply them by $Y - \bar{Y}$. What we are doing here is multiplying the variation of Y by the portion of variation of X_1 *not* explained by X_2. Please see formula 10.1 for how we compute the slope and the intercept.
2. With the computations done in step 1, we compute the partial slope of Y regressed on X_1, holding X_2 constant. What this means is that we regress Y on X_1, as if all the observations of X_2 were the same.
3. We compute an equation regressing X_2 on X_1, and do steps 1–2 again.
4. With the computations done in step 3, we compute the partial slope of Y regressed on X_2, holding X_1 constant.
5. We compute the intercept for the multiple regression equation.
6. We aggregate the multiple regression equation, using the two partial slopes and the intercept.

Let us use a small sample of students for whom we have the following variables: $Y =$ final grade in the semester, $X_1 =$ mean number of hours slept per night, $X_2 =$ mean number of minutes exercised per week. Let us compute the regression equation for the data, where we wish to predict final grade based on mean number of hours slept per night and mean number of minutes exercised per week.

	Final grade	Sleep	Exercise
1	90	6	60
2	70	4	65
3	80	7	75
4	85	6	55
5	50	3	10
6	75	5	55
7	77	5	15
8	60	6	10
9	65	5	15
10	70	7	20
11	60	6	25

We need the information below:

$$\bar{Y} = 71.09, \quad \sum Y = 782, \quad \sum Y^2 = 57{,}004, \quad SS_Y = 1410.91, \quad n = 11$$
$$\bar{X}_1 = 5.4545, \quad \sum X_1 = 60, \quad \sum X_1^2 = 342, \quad SS_{X_1} = 14.73, \quad n = 11$$
$$\bar{X}_2 = 36.82, \quad \sum X_2 = 405, \quad \sum X_2^2 = 21{,}175, \quad SS_{X_2} = 6263.64, \quad n = 11$$
$$\sum X_1{}^* X_2 = 2280$$

We follow the steps outlined above.

1. We regress X_1 on X_2 and obtain the slope b_{12}.With this regression equation we can also predict $\hat{X}_1$ for each value of X_2 and compute the residuals, which are the difference between the observed and expected X_1.

$$SP_{X_1X_2} = 2280 - \frac{60^*405}{11} = 70.91$$

$$b_{X_1X_2} = \frac{SP_{X_1X_2}}{SS_{X_2}} = b_{X_1X_2} = \frac{70.91}{6263.64} = 0.011$$

$$a_{X_1X_2} = \bar{X}_1 - b^*\bar{X}_2 = 5.4545 - 0.011^*36.82 = 5.04$$

Therefore our equation is: $\widehat{X}_1 = 5.04 + 0.011\,(X_2) + \varepsilon$.

With this equation, we do the following:

1. Predict $\widehat{X}_1$.
2. Obtain the residuals.
3. Multiply the residuals by $Y - \bar{Y}$.
4. Sum the products of the latter multiplication.

Obs.	X_2	X_1	$\widehat{X}_1$	$X_1 - \widehat{X}_1$	$Y - \bar{Y}$	$(X_1 - \widehat{X}_1)^*(Y - \bar{Y})$
1	60	6	5.7	0.3	18.91	5.673
2	65	4	5.755	−1.755	−1.09	1.91295
3	75	7	5.865	1.135	8.91	10.111285
4	55	6	5.645	0.355	13.91	4.93805
5	10	3	5.15	−2.15	−21.09	45.3435
6	55	5	5.645	−0.645	3.91	−2.52195
7	15	5	5.205	−0.205	5.91	−1.21155
8	10	6	5.15	0.85	−11.09	−9.4265
9	15	5	5.205	−0.205	−6.09	1.24845
10	20	7	5.26	1.74	−1.09	−1.8966
11	25	6	5.315	0.685	−11.09	−7.59665

$$\sum(X_1 - \widehat{X}_1)^2 = (6 - 5.7)^2 + \ldots (6 - 5.315)^2 = 13.92$$

$$\sum(X_1 - \bar{X}_1)^*(Y - \bar{Y}) = 5.673 + \ldots -7.59665 = 46.57555.$$

2. *The partial slope of Y regressed on X_1, holding X_2 constant is computed.* It is written with a subscript that clarifies which variable is held constant. You can see that the slope multiplies the variation of Y around its mean (namely, $Y - \bar{Y}$) by the portion of the variation of X_1 *not* explained by X_2 (namely, $\sum(X_1 - \widehat{X}_1)^2$), dividing by the latter. This slope will be used for building up the multiple regression equation.

$$b_{Y1.2} = \frac{\sum(X_1 - \hat{X}_1)(Y - \bar{Y})}{\sum(X_1 - \hat{X}_1)^2} = \frac{46.57555}{13.92} = 3.345.$$

3. We compute an equation regressing X_2 on X_1, and do steps 1–2 again.

$$b_{X_2X_1} = \frac{SP_{X_2X_1}}{SS_{X_1}} = b_{X_2X_1} = \frac{70.91}{14.73} = 4.814$$

$$a_{X_2X_1} = \bar{X}_2 - b_{X_2X_1}{}^*\bar{X}_1 = 36.82 - 4.814^*5.4545 = 10.56$$

Therefore our equation is: $\widehat{X}_2 = 10.56 + 4.814\,(X_1) + \varepsilon$.

With this equation, we do the following:

1. Predict $\widehat{X}_2$.
2. Obtain the residuals.
3. Multiply the residuals by $Y - \bar{Y}$.
4. Sum the products of the latter multiplication.

Obs.	X_1	X_2	$\widehat{X}_2$	$X_2 - \widehat{X}_2$	$Y - \bar{Y}$	$(X_2 - \widehat{X}_2)^*(Y - \bar{Y})$
1	6	60	39.444	20.556	18.91	388.71396
2	4	65	29.816	35.184	−1.09	−38.35056
3	7	75	44.258	30.742	8.91	273.91122
4	6	55	39.444	15.556	13.91	216.38396
5	3	10	25.002	−15.002	−21.09	316.39218
6	5	55	34.63	20.37	3.91	79.6467
7	5	15	34.63	−19.63	5.91	−116.0133
8	6	10	39.444	−29.4444	−11.09	326.53396
9	5	15	34.63	−19.63	−6.09	119.5467
10	7	20	44.258	−24.258	−1.09	26.44122
11	6	25	39.444	−14.4444	−11.09	160.18396

$$\sum(X_2 - \widehat{X}_2)^2 = 5922.257$$

$$\sum(X_2 - \widehat{X}_2)^*(Y - \bar{Y}) = 1753.39$$

4. *The partial slope of Y regressed on X_2, holding X_1 constant is below.* It is written with a subscript that clarifies which variable is held constant. You can see that the slope multiplies the variation of Y by the portion of the variation of X_2 *not* explained by X_1, dividing by the latter.

$$b_{Y1.2} = \frac{\sum(X_2 - \hat{X}_2)(Y - \bar{Y})}{\sum(X_2 - \hat{X}_2)^2} = \frac{1753.39}{5922.257} = 0.296.$$

5. *The intercept for the multiple regression equation is*

$$a_{Y.12} = \bar{Y} - b_{Y1.2}\bar{X}_1 - b_{Y.21}\bar{X}_2 = 71.09 - 3.345^*5.4545 - 0.296^*36.82$$
$$= 71.09 - 18.2453 - 10.89872 = 41.946.$$

6. *We ensemble all the equation, using the two partial slopes and the intercept.* The multiple regression is: $\widehat{Y} = 41.946 + 3.345X_1 + 0.296X_2 + \varepsilon$.

The purpose of this exercise was to show the reader that the computation of the multiple regression equation is not terribly difficult or terribly different from the computation of the simple regression equation. Fortunately you will have a highly efficient computer which will do this for you, so let us move on to better things.

11.1.3 An overview of multiple-regression-model building

The process of multiple regression analysis is not unidirectional and it is certainly dictated by the preferences of the individual analyst. This is the way I proceed:

1. Do a brief Y^*X series of plots to see if there are blatant violations of the linearity and homoscedasticity assumptions. Unless the violations are blatant, don't do anything now to transform the raw data because the X with which the Y does not have a linear relation may not even end up in your regression model. Just make a note of which variables may need to be transformed later, should they be incorporated into the model.
2. Do a correlation analysis and note which variables are likely to be good predictors of the Y and which ones suffer from multicollinearity. At this point, which one out of two highly correlated Xs are you more likely to keep in the model?
3. Choose the best model with the raw data. There are several procedures available in both PASW and SAS which allow you to quickly choose the best model. For making this choice you should consider the following "tips":
   a. The principle of parsimony. Keep your model as simple as possible.
   b. High R^2 adjusted.
   c. Minimized MS error.
   d. Significant ANOVA.
   e. Not all the slopes have to be significant. If the model as a whole works well, you may end up having some slopes whose t scores are not significant.
   f. Choose the model with as little multicollinearity as possible (evaluated with tolerance and VIF).
   g. Compute **Mallow's C_p**, a great way to discard models which suffer from lack of fit. The expected value of Mallow's C_p in a non-biased model is the number of estimated parameters in the model (p), which is the number of slopes +1 (the one is added to account for the estimate of the Y-intercept). This is what C_p achieves: by dividing the SS error of the model being considered by an estimate of the variance of the Y and subtracting from this quotient a measure of sample size and the number of parameters estimated from the sample, C_p is able to pick models that are unbiased (which have an unbiased estimate of the true error variance). The variance of Y is estimated with the MS error of the model with all the variables in it. Although other estimates of the variance are possible (Ryan 2009), this is the most common estimate used. The expected value (E) of the SS error in a model with p parameters which fits the data is: $E(\text{SS error}_p) = (n - p)^* s^2$. If the SS error is very different from the expected value, when we plug it in formula 11.2 (shown below) we obtain a number quite different from p, the expected value of C_p. For example, let us look at two different models both with the same number of parameters, where: $s^2 = 827.5166$, $n = 39$, $p = 5$. Since $p = 5$, the expected value of C_p is 5. The first model has a SS error = 27,929 and the second model has a SS error = 35,611. The first C_p is: $C_p = \frac{27{,}929}{827.5166} - (39 - 2^*5) = 4.75$. The second C_p

is: $C_p = \frac{35,611}{827.5166} - (39 - 2^*10) = 14.03$. The first model is superior because its C_p approximates the expected value of 5, while the second one suffers from more lack of fit, with a C_p very much higher than the expected value. Notice that in this case the expected value of the SS error is: E(SS error_p) = (39 – 5)*827.56166 = 28,137.09, which is much closer to 27,929 than to 35,611, the SS error of the model whose C_p was chosen. The reason one model has a much better SS error is that the variables in this model are better predictors of the *Y* than are the variables in the other model. Therefore, if we choose a model with variables that are poor predictors of the *Y*, we will get a model which suffers from lack of fit and which will give us a poor C_p. Mallow's C_p is a great way to discard poor models.

4. Do a full residual analysis of this model. Here you might want to go back and look at the plots of the raw data you did before (step 1). Perhaps there was a slight curve between the *Y* and one of the *X*s which you did not notice before, but which the residuals are making very clear.
5. Transform variables that need to be transformed.
6. Build a new model with the transformed variables and possibly with interaction terms (for example $X^3{}^*X$) and do another residual analysis. Transformations of raw variables are going to be highly correlated with the raw data, so doing a correlation analysis with the transformations (*at this point*) does not make sense.
7. Compare with the first model you computed: were the transformations worth it? Are you sure that your new model is not too complicated? Does it make sense?

Formula 11.2

Formula for the computation of Mallows C_p.

$$C_p = \frac{SS\ error_p}{s^2} - (n - 2^*p)$$

Where $SS\ error_p$ is the SS error of the model being evaluated, s^2 is the MS error of the full model (the model with all the *X*s in it)
n is the sample size in the model being evaluated and
p is the number of parameters $(X + 1)$ in the model being evaluated.

To arrive at the "best models," most computer programs are likely to include only independent variables that have significant t values associated with them. This means that only variables which individually contribute to the explanation of the *Y* will be included. However, it is not a good idea to start the process of model building with such stringent restrictions. Indeed, we need to keep in mind that our primary purpose in doing regression analysis is to explain the variation of *Y*. If our explanatory power increases by adding non-significant variables to the model, and the model makes sense to us, then let it be so. Thus, it is a good idea to change the significance level for the computer to include and to keep variables in the model so that it considers non-significant variables as well. This is what is usually called the "**F to enter and F to remove**." What is critical is that *the F to enter be smaller than the F to remove*. Otherwise you might be unable to remove variables which should be removed from the model when other ones are added. Specific suggestions on the F to enter and F to remove are given below. Most computer packages offer the following procedures (with different names):

1. **All possible equations:** This procedure computes all possible regression equations. If there are p independent variables, there are 2^p equations so that if there are 2 *X*s, there are 4 possible equations but if there are 4 *X*s, then there are 256 possible equations. It doesn't really make sense to use this procedure because you would waste a lot of time looking at poor equations.
2. **Best subset equation:** This procedure determines the best equations with 2, 3, 4, etc. predictors. The criteria used to determine such subsets are C_p , R^2 adj., and the MSE. In my opinion this procedure facilitates choosing the model greatly.
3. **The backward elimination procedure:** This procedure computes a regression equation with all independent variables. Then, the variable that contributes the least to the model is removed and a new model is computed. Once again, the variable that contributes the least is removed and a new model is computed. The process continues until no variable is removed.
4. **The forward selection:** This procedure starts with a simple regression model which includes the variable that best explains the *Y*. This variable is likely to be the *X* that had the highest correlation with the *Y* in the correlation matrix. Then the computer continues to add one variable at a time until all variables which contribute to the explanation of *Y* are in the model. The problem with this procedure is that it may lead to the inclusion of variables that do not make a significant contribution once other independent variables are entered in the regression model. In other words, once a variable is in the model, said variable is not removed even if after the addition of yet another variable the explanatory power of the first one is gone because the second variable is highly correlated with the first one.
5. **The stepwise regression procedure:** This procedure enters and removes variables in turn until the regression equation is satisfactory. It combines the best of forward and backward selection. Eventually, all significant variables should be in the model, and all non-significant variables should not. Please make sure to check the Fs to enter and to remove in your program and make them unusually low so that the program can have the opportunity to enter and remove different variables and to provide you with several possibly good models from which you will choose your final one. In my class I start with setting the F to enter at 0.40 and the F to remove at 0.80. These are just suggestions which you may or may not take.

Since several of these procedures will provide you with the R^2 adj., this is a good moment to take a look at its formula (formula 11.3).

Formula 11.3

Formula for computing the R^2 adj.

$$R^2 \text{ adj.} = 1 - \left[\frac{(n-1)\left(1-R^2\right)}{n-p}\right]$$

Practice problem 11.1

Let us do a complete multiple regression analysis. For space's sake I will not reproduce all of the plots I produced. I will use a computer program to obtain my results, and will not do it by hand (sorry to disappoint you). The data set called "contraception" has several variables which may or may not impact the country's general fertility rate (GFR) such as current percentage use of contraception, the median level of education for males and females, the median age at marriage and at first birth, and the percent of live births in the three years preceding the survey receiving no antenatal care during pregnancy. Let us use these variables to construct a multiple regression model where GFR is the Y. We follow the steps outlined above.

1. *Do a brief Y*X series of plots to see if there are blatant violations of the linearity assumption.* The plots of GFR against contraception, both education variables and age at marriage and first birth look quite linear and negative. The plot of GFR against antenatal care suggests non-homogeneous variance: as the X increases so does the variance of the Y. In addition, the line between them seems pretty flat. I would not be surprised if antenatal care is not chosen as a predictor.
2. *Do a correlation analysis.* Table 11.1 shows the correlation matrix among all the variables. The matrix is arranged so that the correlation coefficient between two variables is shown first (the correlation between contraception and GFR is −0.75616), followed by the p value associated with the coefficient ($p < 0.0001$), followed by the sample size. The differences in sample sizes are due to the missing values of the data set. The dependent variable (GFR) is highly correlated with all variables except "no antenatal care." However, the p value associated with the correlation between these two variables is 0.06. Therefore, I would not assume at this point that "no antenatal care" has nothing to contribute to the model. We may have a problem with multicollinearity since all of the variables are related to some others. We are probably going to have to drop one of the education variables, for example.
3. *Choose the best model with the raw data.* Below are the best models considering the MSE, the R^2 adj., and C_p. While the first model has the highest R^2 adj. and the smallest MSE, its C_p is not good, since C_p should be about 5 (the number of Xs in the model +1). Indeed, the C_ps for the second and third model are not good either since they should be about 6. Model #4 looks good in terms of its C_p which approaches

Table 11.1 Correlation matrix of the general fertility rate data.

	Contraception	GFR	Male education	Female education	First marriage	First birth	No antenatal
Contraception	1.00000						
	44						
GFR	−0.75616	1.00000					
	<0.0001						
	44	44					
Male ed.	0.72090	−0.71498	1.00000				
	<0.0001	<0.0001					
	41	41	41				
Females ed.	0.70709	−0.71430	0.91912	1.00000			
	<0.0001	<0.0001	<0.0001				
	41	41	41	41			
First marriage	0.49144	−0.57960	0.70520	0.72242	1.00000		
	0.0009	<0.0001	<0.0001	<0.0001			
	42	42	39	39	42		
First birth	0.46456	−0.69951	0.62980	0.60868	0.85392	1.00000	
0.0015	<0.0001	<0.0001	<0.0001	<0.0001			
	44	44	41	41	42	44	
No antenatal	−0.35198	0.28509	−0.41653	−0.47334	−0.57089	−0.33130	1.00000
	0.0223	0.0672	0.0067	0.0018	0.0001	0.0321	
	42	42	41		41	40	42

the desired value of 5. Model #6 is parsimonious, with only three variables and with an adjusted R^2 of 0.7490. Model #7 is the most parsimonious (only two Xs) and has an adjusted R^2 of 0.7264, a perfect C_p, and a MSE comparable with that of model #4 but higher than that of model #6. Therefore at this point the best models appear to be the 4th, the 6th, and 7th. We take a look at them.

Number of variables in model		Adjusted R^2	R^2	C_p	MSE	Variables in model
1.	4	0.7300	0.7585	3.0079	779.07495	*
2.	5	0.7219	0.7585	5.0035	802.57222	**
3.	5	0.7219	0.7585	5.0051	802.61219	***
4.	4	0.7154	0.7453	4.7490	821.45141	****
5.	6	0.7132	0.7585	7.0000	827.56166	*****
6.	3	0.7490	0.7292	4.0000	808.30270	******
7.	2	0.7264	0.7131	3	821.22394	*******

* Contraception, females' education, age at first marriage, age at first birth.
** Contraception, males' education, females' education, age at first marriage, age at first birth.
*** Contraception, females' education, age at first marriage, age at first birth, no antenatal care.
**** Contraception, males' education, age at first marriage, age at first birth.
***** Contraception, males' education, females' education, age at first marriage, age at first birth.
****** Contraception, age at first marriage, age at first birth.
******* Contraception, age at first birth.

Model #4: You can see that the entire model is significant below the 0.001 level. You can also see that none of the tolerance or VIF values is of concern, so our model does not suffer from multicollinearity. One of the variables (male education) itself does not make a significant contribution.

		Analysis of variance			
Source	df	Sum of squares	Mean square	F value	p
Model	4	81,736	20,434	24.88	<0.0001
Error	34	27,929	821.45141		
Total	38	109,665			
R^2	0.7453.	R^2 adj.	0.7154		

			Parameter estimates					
Variable	df	Parameter estimate	Standard error	t	df	p	TOL	VIF
Intercept	1	461.8297	72.49644	6.37	1	<0.0001	.	0
Contraception	1	−1.43010	0.40814	−3.501	1	0.0013	0.47074	2.12429
Male education	1	−4.25517	3.05247	−1.39	1	0.1724	0.29826	3.35283
Age at first marriage	1	10.64874	5.26902	2.02	1	0.0512	0.22355	4.47334
Age at first birth	1	−22.13802	5.54250	−3.99	1	0.0003	0.26322	3.79909

The model equation is written as:

$$\widehat{GFR} = 451.8297 - 1.43\,(\textit{contraception}) - 4.25\,(\textit{male education})$$
$$+\,10.64\,(\textit{age at first marriage}) - 22.13(\textit{age at first birth}).$$

The model indicates that if we increase contraception by one unit we decrease GFR by −1.43, that if we increase male education by one unit we decrease the GFR by 4.25, that if we increase age at first marriage by one unit we increase the GFR by 10.64, and that if we increase the age at first birth we decrease the GFR by 22.13.

Model #6: You can see that the entire model is significant below the 0.001 level. You can also see that none of the tolerance or VIF values is of concern, so our model does not suffer from multicollinearity. It is interesting that after we removed male education, age at first marriage is not significant. This shows the importance of re-evaluating the contributions of variables in the model after we remove other ones.

		Analysis of variance			
Source	df	Sum of squares	Mean square	F value	p
Model	3	91,647	30,549	37.79	<0.0001
Error	38	30,716	808.30270		
Total	41	122,362			
R^2	0.7490	R^2 adj.	0.7292		

		Parameter estimates						
Variable	df	Parameter estimate	Standard error	t	df	p	TOL	VIF
Intercept	1	519.28569	60.68249	8.56	1	<0.0001.	0	
Contraception	1	−1.84414	0.30504	−6.05	1	<0.0001	0.74583	1.34078
Age at first marriage	1	8.06451	4.66235	1.73	1	0.0918	0.26186	3.81878
Age at first birth	1	−22.81583	5.33923	−4.27	1	0.0001	0.26630	3.75523

The model equation is written as:

$$\widehat{Y} = 519.28 - 1.84\,(\mathit{contraception}) + 8.06\,(\mathit{age\ at\ first\ marriage}) - 22.81(\mathit{age\ at\ first\ birth}).$$

The model indicates that as we increase contraception by one unit we decrease GFR by 1.84, as we increase age at first marriage by one unit we increase GFR by 8.06, and as we increase age at first birth by one unit we decrease GFR by 22.81.

***Model** #7:* You can see that the entire model is significant below the 0.001 level. You can also see that none of the tolerance or VIF values is of concern, so our model does not suffer from multicollinearity. This is certainly my favorite model because of its parsimony. Thus, this is the model that I choose as the best.

		Analysis of variance			
Source	df	Sum of squares	Mean square	F value	$p > F$
Model	2	89,397	44,699	54.43	<0.0001
Error	41	33,670	821.22394		
Total	43	123,068			
R^2	0.7264	R^2 adj.	0.7131		

		Parameter estimates						
Variable	df	Parameter estimate	Standard error	t	df	p	TOL	VIF
Intercept	1	507.81028	59.93798	8.47	1	<0.0001		
Age at first birth	1	−14.92132	3.09962	−4.81	1	<0.0001	0.78418	1.27521
Contraception	1	−1.76524	0.29614	−5.96	1	<0.0001	0.78418	1.27521

The model equation is written as: $\widehat{Y} = 507 - 14.92(\mathit{age\ at\ first\ birth}) - 1.75$ $(\mathit{contraception})$. This model indicates that as we increase age at first birth by one unit we expect the GFR to decrease by 14.92, and as we increase the use of contraception by one unit we expect to decrease the GFR by 1.75.

Table 11.2 Descriptive statistics of the general fertility rate for each region.

Region	n	Mean	Std. Dev.
Sub-Saharan Africa (0)	25	183.88	34.77
North Africa, West Asia, and Europe (1)	6	7	31.68
South and Southeast Asia (2)	7	110	14.17
Latin America and Caribbean (3)	6	104	16.99

4. *Do a full residual analysis of this model* (*model #7*). Quite fortunately I did not detect any problems with the residuals. The scatter plot of residuals against the two *X*s was random, the histogram of the residuals was quite normal, and the normality test was not significant. I do not see any reason to transform the independent variables. Therefore we do not need to do steps 5, 6, and 7 outlined at the beginning of this section.

11.1.4 Dummy independent variables

I hope you recall that in ANOVA the independent variables are not continuous. Rather, they are either dichotomous (in which case we deal with the problem as a t test) or they have a few discrete values (in which case we have an ANOVA). Let us recall that ANOVA is only a form of regression analysis in which the dependent variable is continuous numeric and the independent variable is discontinuous. So far in this chapter we have looked at regression analysis with models in which the Y and the Xs are both continuous. However, regression analysis can incorporate discontinuous independent variables just as if we were doing an ANOVA. The purpose of analyzing a data set in this manner instead of using an ANOVA is still to explain the variation of the Y and to be able to predict the Y. However, we are not going to be making hypotheses tests about the means of the subgroups. Rather, we see subgroup membership as an independent variable which may (or may not) have an effect on the dependent variable.

Table 11.2 shows descriptive statistics of the general fertility rate for the four regions from which we have data. You will notice that while the sample size of Africa is adequate ($n = 25$), the sample sizes for the other regions are very small ($n = 6$ and $n = 7$). Therefore, it would be very difficult to defend a decision to submit this data to an analysis of variance which assumes normality of each subsample with such small groups. Instead, it is much better to ask if group membership significantly affects the behavior of the GFR.

Let us look at a regression equation written in non-parametric notation: $\hat{Y} = a + b_1X_1 + b_2D$, where X_1 is a continuous independent variable and D is the dummy variable, where D can take values of 0 or 1. This equation will take two forms, depending on the value of D. If $D = 1$ and b_2 is the slope of the dummy variable, then: $\hat{Y} = a + b_1X_1 + b_21 = (a + b_2) + b_1X_1$. In contrast, if $D = 0$ and b_2 is the slope of the dummy variable, then: $\hat{Y} = a + b_1X_1 + b_20 = a + b_1X_1$. These two equations show that we will be working with two lines, depending on the value taken by D. If $D = 0$ then the intercept is a, and the slope of the line is determined by b_1. But if $D = 1$ then the intercept is $a + b_2$, and the slope of the line is determined by b_1. If D has 3 values, then there will be 3 lines, etc. It is customary to have 2 or three dummy variable levels to keep the interpretation of the equation lines simple.

Let us think of a two-way ANOVA design for a moment. If we test for equality of years of formal education between male and female students who belong to three different religious groups and we find significant interaction, we cannot continue our analysis: one gender gets more formal education in one of the groups while the other gender gets more education in the other groups. In such a case we can do our analysis with a regression model in which we enter both religion and gender as independent variables and add another variable for their interaction, namely, $X_1{}^*X_2$. By doing our study in a regression analysis rather than a two-way ANOVA approach we have switched the focus from testing if the mean number of years of formal education differs among the subgroups to understanding how the independent variables affect the Y. Please note that I am not saying that an interaction term will always be necessary, but that it is something worth trying in a situation such as this. The beauty of a regression analysis approach to this problem is that we would obtain two different regression lines, in which we could see the effect of the other independent variable (such as religion) on the Y, while controlling for gender.

Please note that when you are doing regression analysis with a dummy variable you do not gain any information by plotting the raw data or the residuals against the dummy: you will only get two or three clumps of data. Of course, you should check for linearity in the raw data and randomness in the residuals with the continuous independent variables.

Practice problem 11.2

Let us continue our analysis of the general fertility rate, by adding a dummy variable representing region. Let us also confine our analysis to two regions, namely, region 0 (Africa) and region 1 (North Africa, West Asia, and Europe). Because Africa is the reference point (it is represented by a 0) then the slope associated with the variable "region" will tell us how much region one is expected to differ from Africa. The data were entered as follows (only a few lines are reproduced here). Please note that the dummy variable is entered as a single column.

Region ID	Country ID	Country	Year	Contraception	General Fertility Rate (GFR)	Male education
0	1	Benin	2006	17	204	2
1	27	Armenia	2005	33	58	7
1	28	Azerbaijan	2006	32	66	10

I performed a regression analysis keeping my favorite model (which included age at first birth and contraception), and added the dummy variable (which has two levels: 0 = Africa, 1 = North Africa, West Asia, and Europe). The new model is shown below:

		Analysis of variance of model incorporating dummy variable			
Source	df	Sum of squares	Mean square	F value	p
Model	3	68,036	22,679	25.07	<0.0001
Error	27	24,426	904.64999		
Corrected total	30	92,461			
R^2	0.7358	R^2 adj.	0.7065		

			Parameter estimates			
Variable	df	Parameter estimate	Standard error	t	df	p
Intercept	1	416.51895	110.25658	3.78	1	0.0008
Region ID	1	–48.00065	24.33399	–1.97	1	0.0589
Contraception	1	–1.07264	0.46269	–2.32	1	0.0282
Age at first birth	1	–10.68286	5.67806	–1.88	1	0.0707

The equation of this model is written as

$$\hat{Y} = 416 - 48/(region) - 1.07/(age\ at\ first\ birth) - 10/(contraception) + \varepsilon$$

According to this model, a change from Africa to Europe/North Africa/West Asia is predicted to decrease the GFR by 48 units. The meaning of the other two slopes remains the same.

Let us look at how we use this equation to predict values. The first observation has the following variates: region = 0, contraception = 17, GFR = 204, and age at first birth = 20. If we substitute the observed values we obtain:

$$\hat{Y} = 416 - 48\,(0) - 1.07\,(17) - 10.68\,(20) = 416 - 18.19 - 213.6$$
$$= 184.21. \text{ The residual is } 204 - 184.62 = 19.79.$$

The second observation has the following variates: region = 1, contraception = 33, GFR = 58, and age at first birth = 22. If we substitute the observed values we obtain: $\hat{Y} = 416 - 48\,(1) - 1.07\,(33) - 10.68\,(22) = 416 - 48 - 35.31 - 234.96 = 97.73$. The residual is 58–97.73 = 39.73. Thus, the second observation is predicted to have a much lower GFR than the first observation because it is from region one, where the GFR is much lower.

What about if North Africa/Europe/West Asia had been assigned the value of 0 and Africa the value of 1? Then the slope for region would have been 48.00, indicating that as we go from region 0 to region 1 the GFR should increase 48 units. Therefore, the interpretation of the slope is simple enough; you just have to remember what category of your dummy variable was assigned a value of 0, and consider whether the slope is negative or positive.

We now do this analysis with three levels of the dummy variable and include South/South east Asia. The dummy variable was now entered in three different columns (each named for a region), so that if an observation is from a region then it has a "1" under that region but "0" for the other ones. Part of the data set is reproduced, so that you can see how the data were entered. Only two observations per regions are shown below.

Africa	Europe	Asia	Contraception	GFR	Age at first birth
1	0	0	17	204	19
1	0	0	26	177	18
0	1	0	33	58	21
0	1	0	32	66	22
0	0	1	56	107	15
0	0	1	24	111	20

The regression analysis with the three levels of the dummy yields the following results:

		Analysis of Variance			
Source	df	Sum of squares	Mean square	Value	p
Model	4	89,370	22,343	25.06	<0.0001
Error	37	32,992	891.67659		
Total	41	122,362			
R^2	0.7304.	R^2 adj.	0.7012		

		Parameter estimates			
Variable	df	Parameter estimate	Standard error	t	p
Intercept	1	339.40238	55.66717	6.10	<0.0001
Age at first birth	1	−7.07506	3.13386	−2.26	0.0300
Contraception	1	−1.43256	0.35571	−4.03	0.0003
Asia	1	−39.68299	13.76726	−2.88	0.0065
Europe	1	−48.07814	15.86603	−3.03	0.0044

The equation for Africa is:

$$\widehat{Y} = 339.40238 - 7.07506\,(\textit{age at first birth}) - 1.43256\,(\textit{contraception}) + \varepsilon.$$

The equation for Asia is:

$$\widehat{Y} = 339.40238 - 7.07506\,(\textit{age at first birth}) - 1.43256\,(\textit{contraception}) - 39.6829 + \varepsilon.$$

The equation for Europe is:

$$\widehat{Y} = 339.40238 - 7.07506\,(\textit{age at first birth}) - 1.43256\,(\textit{contraception}) - 48.0781 + \varepsilon.$$

Therefore, if we change the region from Africa to Asia (keeping age at first birth and contraception constant), we expect to see a decrease in the GFR of almost 40 units (39.68). If we change the region from Africa to Europe (keeping the other two independent variables constant), we expect to see a decrease in the GFR of 48 units. Therefore, you can see that the slopes can be used to understand the effect of the independent variables on the dependent variable when we have dummy variables, just as if we were working with continuous independent variables. The predicted values are computed as before.

11.2 An overview of logistic regression

Logistic regression is a statistical technique that is becoming more and more frequently used in the anthropological literature. What is unique to logistic regression is that the dependent variable in the model is dichotomous. Therefore, we do not have an assumption of normality of the *Y*, nor do we have any reason to check for linearity of the relationship between the *Y* and the *X*s or of normality of the residuals. The purpose of logistic regression is still prediction of the *Y*, given a series of independent variables. In logistic regression, we are asking questions such as:

1. If we have data on a Senator's party affiliation, his/her gender, his/her district's median income, can we predict if she or he will vote yes (0) or no (1) on a particular bill?
2. If we have implemented a program to prevent teen pregnancy, and we have data on male teen's attendance to the program, hours each week exercising, hours each week studying, attendance at school, household income, household composition, can we predict if he will use a condom (yes = 0) or not (no = 1)?
3. If we have data on pregnant women's socioeconomic status, attendance of a birthing course, whether the woman has other children, whether the woman has a current partner, the woman's education, and the woman's job status, can we predict if the woman will breastfeed (0) or not (1)?

What all of these research projects have in common is that the dependent variable can take two forms: yes or no, 0 or 1. Thus, logistic regression is very popular in epidemiological studies, in which the dependent variable may be mortality (will the patient survive the procedure?) or morbidity (will the patient develop cancer after 10 years?). The keen observer may have noticed that in some studies the independent variables and the dependent variable could have been collected retrospectively (such as in example 1; we can do a historical study on voting records). At the same time, we can do a study which takes months or years collecting the independent variables and finish collecting the dependent variable by a certain date (for example, we can begin the study with all women who start attending a prenatal clinic on January 1 of a year, and conclude our study on December 31, and consider only those women who delivered by the latter date). Logistic regression is applicable in both cases because what we want to do is to predict the response variable even if it has already occurred (Kleinbaum and Klein 2010). Thus, with logistic regression we are still trying to understand and predict the dependent variable with a model which usually includes more than one independent variable. The model may include both continuous and dummy variables, and it may be chosen using the procedures discussed above, such as stepwise, forward, etc.

However, the fact that the dependent variable is binary requires that in most cases the estimates of the parameters be computed with a maximum likelihood instead of a least-squares approach. Logistic regression also involves much more complex methods for evaluating goodness of fit. Our good-old R^2 is just not going to cut it. For that reason, this section will help you understand logistic regression at its most fundamental level, but it will not prepare you to do far-reaching logistic regression on your own. Indeed, in second-level regression analysis textbooks, logistic regression is usually reviewed in passing as I am doing here. Instead, logistic regression is usually covered in its own specialized set of texts, which you are encouraged to consult for your own edification (Hosmer and Lemeshow 2000; Jaccard 2001; Kleinbaum and Klein 2010; Pampel 2000).

Let us recall that a parametric multiple regression model looks like $Y = \alpha + \beta_1 + \ldots \beta_k$ where k is the number of independent variables (which may include dummy variables) and the intercept and the slopes are as discussed above. In a logistic regression model we cannot express our model in this manner because the Y can only take two values: 0 or 1. Therefore, we can only predict success or failure, death or survival, 0 or 1, and so forth. When we predict that a congressman will vote one way instead of another, we are really talking about the odds of the congressman voting in one way rather than the other. That is why the odds ratio is frequently discussed in conjunction with logistic regression.

How are we to write down the dependent variable, then? Let us take a brief detour before we answer this question. If previous studies indicate that the probability that a woman will breastfeed upon discharge from the hospital is $p = 0.7$ and the probability that she will not is $1 - p = 0.3$, then the odds that she will breastfeed are: $\frac{p}{1-p} = \frac{0.7}{0.3} = 2.33$. We can of course reverse this and say that the odds that she will not breastfeed are $\frac{0.3}{0.7} = 0.428$. I don't know about you, but many people find the fact that these two ratios are not symmetrical quite uncomfortable. To ease you and me out of such discomfort, let us do the following: $\ln\left(\frac{0.7}{0.3}\right) = 0.8473$ and $\ln\left(\frac{0.3}{0.7}\right) = -0.8473$. In logistic regression the *dependent variable is the natural log* (the **logit**) *of the odds ratio* because with the natural log we can compute symmetrical confidence limits. The confidence limits are of crucial importance to the interpretation of the odds ratios for reasons you learned in chapter eight: if an odds ratio includes the value of 1 in its confidence interval then it is not significant. Thus, with logistic regression we can predict the log odds that a woman will breastfeed, and then provide an odds ratio with a confidence interval which will allow us to decide if the odds is significant.

Formula 11.4

Formula for the computation of the logistic regression model.

$$\log\left(\frac{\pi}{1-\pi}\right) = \beta_0 + \beta_1 + \ldots \beta_k, \quad \text{where there are } k \text{ independent variables.}$$

Let us do a logistic regression using the data set called "Cranial morphology." I asked SAS to choose the best set of independent variables (in a stepwise manner) to predict if a cranium is a male or a female. I did not change the default F to enter and F to remove, in the hope that you would get the same output that I did. The final model is:

Parameter	df	Estimate	Standard error	Chi-square	p
Intercept	1	35.2628	8.4610	17.3695	<0.0001
BBH	1	−0.1375	0.0499	7.5887	0.0059
NLH	1	−0.3168	0.1181	7.2003	0.0073

We write the equation as usual:

$$\log\left(\frac{f}{m}\right) = 35.2628 - 0.1375\,(BBH) - 0.3168\,(NLH) + \varepsilon$$

We can interpret the slopes as follows: for every unit increase in BBH we estimate the log odds of a cranium being a female to be decreased by –0.1375 while holding the other X fixed; and for every unit increase in NLH we estimate the log odds of a cranium being a female to decrease by –0.3168 while holding the other X fixed. The p value associated with both slopes is significant. How do I know that the reference point is the female and not the male crania? Since I obtained these estimates using SAS, my raw data actually had Fs and Ms in the Excel spreadsheet, and in its log SAS had the courtesy of letting me know the following: *"PROC LOGISTIC is modeling the probability that Sex = 'F'."*

Now we show the odds ratios (the really neat part!) and their confidence limits.

Effect	Point estimate	Confidence	Limits
BBH	0.871	0.790	0.961
NLH	0.728	0.578	0.918

If I exponentiate the slope for BBH then I get the odds ratio point estimate provided by SAS, namely: for BBH: $e^{-0.1375} = 0.871$, and for NLH: $e^{-0.3168} = 0.728$. The interpretation of these odds ratios is basically the same as that of the slopes: for one unit increase in BBH, the odds of the cranium being a female is 0.871, and for every unit increase in NLH, the odds of the cranium being a female is 0.728. Note that the confidence limits do not include 1, so that the odds are significant.

Practice problem 11.3

I changed the data set analyzed in the previous paragraphs so that males = 0 and females = 1 and I obtained the following logistic model:

Parameter	df	Estimate	Standard error	Wald Chi-square	$p >$ Chi-square
Intercept	1	−32.0056	6.7755	22.3133	<0.0001
BBH	1	0.1843	0.0458	16.2013	<0.0001
NLH	1	0.1295	0.0639	4.1012	0.0429

From this output we obtain the information for writing down the equation:

$$\log\left(\frac{m}{f}\right) = -32.0056 + 0.1843\,(BBH) + 0.1295\,(NLH) + \varepsilon$$

We can interpret the slopes as follows: for every unit increase in BBH we estimate the log odds of a cranium being a male to be increased by 0.1843 while holding the other X constant and for every unit increase in NLH we estimate the log odds of a cranium being a male to be increased by 0.1295 while holding the other X constant. Now to the odds ratios, which are shown below with their confidence limits.

Effect	Point estimate	Confidence	Limits
BBH	1.202	1.099	1.315
NLH	1.138	1.004	1.290

If I exponentiate the slope for BBH then I get the odds ratio point estimate provided by SAS, namely: for BBH: $e^{0.1843} = 1.202$, and for NLH: $e^{0.1295} = 1.138$. The interpretation of these odds ratios is basically the same as that of the slopes: for one unit increase in BBH, the odds of the cranium being a male is 1.202, and for one unit increase in NLH, the odds of the cranium being a male is 1.138. Let us look at the confidence limit. If the limit includes the value of 1 then it is not significantly different from 1. You can see that the lower confidence limit for NLH is dangerously close to 1, so I would take the ability of this variable to differentiate male and female crania with a grain of salt.

11.3 Writing up your results

If you have performed a full regression analysis, including the initial inspection of the scatter plots, a correlation matrix analysis, choosing of the model by applying different procedures, evaluating competing models, doing a residual analysis, transforming variables, re-doing the regression model and re-evaluating the model, you most certainly want people to know that you have done a lot of work. But readers do not want to hear about it. Readers want to know that you have not violated the assumptions, that you evaluated the residuals, and that you have good reasons for choosing your final model. When it comes to writing a thesis, you may have more space to include details about the various iterations you went through. But when you are writing a research paper you must be brief. Indeed, if you had to transform your data perhaps all you need to say is that your model incorporated (say) $\sqrt{X_1}$ instead of the raw variable itself. This is a sure way to tell the reader that you had to transform the data, which shows that you were a conscientious analyst. Believe me, the reader is not interested in how much trouble you had when doing the regression analysis.

Logistic regression in anthropology

I think everybody agrees that there exists a serious childhood obesity problem in the United States, and that it is desirable to understand possible causes of the problem. In a carefully controlled study, Metzger and McDade (2010) are able to show that the odds that a child develops obesity are significantly impacted by whether the child was breastfed or not. What is particularly strong about this study is that the authors were able to obtain data on siblings that were and were not breastfed. The authors show that breastfed infants are 41% less likely than non-breastfed infants to become overweight or obese (Metzger and McDade 2010).

11.4 Chapter 11 key concepts

- Review the differences between independent and dependent variables.
- What types of variables (continuous, discontinuous, etc.) are used in multiple regression analysis as dependent and independent variables?
- What makes a variable a dummy variable?
- How is multiple regression analysis similar to ANOVA?
- Assumptions of regression analysis.
- When are transformations advisable?
- Review the different steps in regression analysis.
- How should models be chosen?
- When is logistic regression appropriate?
- Discuss the dependent variable in logistic regression.
- Review the model-building procedures available in the computer package you choose.

11.5 Computer resources

- PASW does not allow for the computation of Mallows CP as part of its menu-based interface. SAS does, but you need to click on selection method in the "additional options" menu in the linear regression window.
- SAS will provide you with the best subset regressions, chosen by CP or by minimized MSE or by highest R^2adj. However, it does not find all of the possible best equations. Indeed, the final equation I chose in this chapter I found on my own.

11.6 Chapter 11 exercises

All data sets are available for download at www.cambridge.org/9780521147088.

- Use the data set called "menarche" and do a multiple regression analysis where $Y =$ number of surviving children.
- Use the data set we analyzed in this chapter but make the $Y =$ use of contraception. What is the best model for prediction of use of contraception?
- Use the data set called "school snacks" to do a multiple regression analysis where $Y =$ student height. This data set has two different dummy Xs. What is your best model?
- Use the data set called "seasonal workers." Do a multiple regression analysis were $Y =$ SBP.
- Use the data set called "seasonal workers." Do a multiple regression analysis were $Y =$ DBP.
- Use the data set called "school snacks" and do two logistic regression analyses. The first one with $Y =$ activity level, the second one with $Y =$ snack machines present.

References

Aldrich, J. (1997) R. A. Fisher and the making of maximum likelihood 1912–1922. *Statistical Science*, **12**, 162–176.

Alvarez, G. F., C. Ceballos & C. Quinteiro (2009) The role of inbreeding in the extinction of a European royal dynasty. *PloS One*, **4**, 7.

Arnold, W. H., V. A. Fedorischeva & N. I. Yabluchansky (2008) Craniometric measurements of artificial cranial deformations in Eastern European skulls. *Anthropologischer Anzeiger*, **66** (2), 139–146.

Bayes, T. (1991) An essay towards solving a problem in the doctrine of chances. *M D Computing*, **8**, 157–171.

Bertorelle, G., A. Benazzo & S. Mona (2010) ABC as a flexible framework to estimate demography over space and time: some cons, many pros. *Molecular Ecology*, **19**, 2609–2625.

Bohrnstedt G. & D. Knoke (1988) *Statistics for Social Data Analysis*. Prospect Heights, IL: F. E. Peacock Publishers.

Bramblett, C. (1994) *Patterns of Primate Behavior*. Prospect Heights, IL: Waveland Press.

Calonge, N., D. B. Petitti, T. G. DeWitt *et al.* (2009) Screening for breast cancer: US Preventive Services Task Force Recommendation Statement. *Annals of Internal Medicine*, **151**, 716–236.

Cohen, J. (1988) *Statistical Power Analysis for the Behavioral Sciences*. Hillsdale, NJ: Lawrence Erlbaum.

Cohen, J. (1992) A power primer. *Psychological Bulletin*, **112**, 155–159.

De Greef, S., P. Claes, W. Mollemans et al. (2005) Semi-automated ultrasound facial soft tissue depth registration: method and validation. *Journal of Forensic Sciences*, **50** (6), 1282–1288.

Dielman, T. (2005) *Applied Regression Analysis. A Second Course in Business and Economic Statistics*. Belmont, CA: Brooks/Cole Thomson Learning.

Draper, N. & H. Smith (1998) *Applied Regression Analysis*. New York, NY: John Wiley & Sons.

Ellison, P. T., R. G. Bribiescas, G. R. Bentley *et al.* (2002) Population variation in age-related decline in male salivary testosterone. *Human Reproduction*, **17**, 3251–3253.

Fienberg S. & N. Lazar (2001) William Sealy Gosset. In C. Heyde & E. Seneta (Eds.), *Statisticians of the Centuries*. New York, NY: Springer.

Fisher, R. (1993) *Statististical Methods, Experimental Design, and Scientific Inference. A Re-issue of Statistical Methods for Research Workers, the Design of Experiments and Statistical Methods and Scientific Inference*. Oxford: Oxford University Press.

Fleiss, J., B. Levin & M. Paik (2003) *Statistical Methods for Rates & Proportion*. Hoboken, NJ: John Wiley & Sons.

Futuyma, D. (1995) *Science on Trial: The Case for Evolution*. Sunderland, MA: Sinauer Associates.

Gould, S. (1993) Cordelia's dilemma. *Natural History*, **2**, 10–18.

Greif, J. M. (2010) Mammographic screening for breast cancer: an invited review of the benefits and costs. *Breast*, **19**, 268–272.

Griffith, D. A. & R. Haining (2006) Beyond mule kicks: the Poisson distribution in geographical analysis. *Geographical Analysis*, **38**, 123–139.

Hocking, R. (2003) *Methods and Applications of Linear Models. Regression and the Analysis of Variance*. New York, NY: John Wiley & Sons.

Hosmer, D. W. & S. Lemeshow (2000) *Applied Logistic Regression*. New York, NY: John Wiley & Sons.

Jaccard, J. (2001) *Interaction Effects in Logistic Regression*. Thousand Oaks, CA: Sage.

Jenkins, G. M. (1979) *Practical Experiences with Modelling and Forecasting Time Series*. Channel Islands: GJP Ltd.

Kisioglu, A. N., A. R. Ormeci, E. Uskun, M. Ozturk & K. Ongel (2010) Effects of a formal training programme on consanguineous marriages on high school students' knowledge and attitudes: an interventional study from Turkey. *Journal of Biosocial Science*, **42**, 161–176.

Kleinbaum, D. & M. Klein (2010) *Logistic Regression. A Self-learning Text*. New York, NY: Springer.

Kohlrausch, F. B., S. M. Callegari-Jacques, L. T. Tsuneto et al. (2005) Geography influences microsatellite polymorphism diversity in Amerindians. *American Journal of Physical Anthropology*, **126**, 463–470.

Kohrt, B. A. & C. M. Worthman (2009) Gender and anxiety in Nepal: the role of social support, stressful life events, and structural violence. *CNS Neuroscience & Therapeutics*, **15**, 237–248.

Krishan, K., T. Kanchan & J. A. DiMaggio (2010) A study of limb asymmetry and its effect on estimation of stature in forensic case work. *Forensic Science International*, **200**, 181.e1–181.e5.

Langley-Shirley, N. & R. L. Jantz (2010) A Bayesian approach to age estimation in modern Americans from the clavicle. *Journal of Forensic Sciences*, **55**, 571–583.

Lisker, R., E. Ramirez & V. Babinsky (1996) Genetic structure of autochthonous populations of MesoAmerica: Mexico. *Human Biology*, **68**, 395–404.

Madrigal, L. (1989) Hemoglobin genotype, fertility, and the malarial hypothesis. *Human Biology*, **61**, 311–325.

Madrigal, L. & T. Koertvelyessy (2003) Epidemic cycles in agricultural populations: a cross-cultural study. *Human Biology*, **75**, 345–354.

Madrigal, L., J. Brady, M. Raxter, E. Ruiz, F. Otarola & M. Blell (2011) Obesity, hypertension and migration: a meta analysis of populations of the South Asian diaspora. *Human Biology*, **83**, 71–86.

Manly, B. F. J. (1992) *The Design and Analysis of Research Studies*. Cambridge: Cambridge University Press.

Martin, S. A., D. Guatelli-Steinberg, P. W. Sciulli & P. L. Walker (2008) Brief communication: comparison of methods for estimating chronological age at linear enamel formation on anterior dentition. *American Journal of Physical Anthropology*, **135**, 362–365.

Mendenhall, W. & T. Sincich (2003) *A Second Course in Statistics: Regression Analysis*. Upper Saddle River, NJ: Prentice Hall.

Metzger, M. W. & T. W. McDade (2010) Breastfeeding as obesity prevention in the United States: a sibling difference model. *American Journal of Human Biology*, **22**, 291–296.

Myung, I. J. (2003) Tutorial on maximum likelihood estimation. *Journal of Mathematical Psychology*, **47**, 90–100.

Neyman, J. & E. Pearson (1933) On the problem of the most efficient tests of statistical hypotheses. *Philosophical Transactions of the Royal Society of London Series A - Containing Papers of a Mathematical or Physical Character*, **231**, 289–337.

Pagano, M. & K. Gauvreau (1993) *Principles of Biostatistics*. Belmont, CA: Wadsworth Inc.

Pakhretia, S. & R. S. Pirta (2010) A behavioural study of the sheep and goats of the transhumant gaddis. *Journal of Human Ecology*, **29**, 93–100.

Pampel, F. C. (2000) *Logistic Regression: A Primer*. Thousand Oaks, CA: Sage.

Pollard, W. (1986) *Bayesian Statistics for Evaluation Research*. Beverly Hills, CA: Sage.

Rosner, B. (2006) *Fundamentals of Biostatistics*. Duxbury: Thomson Brooks/Cole.

Ryan, T. P. (2009) *Modern Regression Methods*. Hoboken, NJ: John Wiley & Sons.

Sartorius, B., H. Jacobsen, A. Torner & J. Giesecke (2006) Description of a new all cause mortality surveillance system in Sweden as a warning system using threshold detection algorithms. *European Journal of Epidemiology*, **21**, 181–189.

Scott, R. M. & H. R. Buckley (2010) Biocultural interpretations of trauma in two prehistoric Pacific Island populations from Papua New Guinea and the Solomon Islands. *American Journal of Physical Anthropology*, **142**, 509–518.

Sokal, R. & F. Rohlf (1995) *Biometry*. New York, NY: W. H. Freeman and Company.

Steadman, D. W., B. J. Adams & L. W. Konigsberg (2006) Statistical basis for positive identification in forensic anthropology. *American Journal of Physical Anthropology*, **131**, 15–26.

Thevissen, P. W., S. Fieuws & G. Willems (2010) Human dental age estimation using third molar developmental stages: does a Bayesian approach outperform regression models to discriminate between juveniles and adults? *International Journal of Legal Medicine*, **124**, 35–42.

Towner, M. C. & B. Luttbeg (2007) Alternative statistical approaches to the use of data as evidence for hypotheses in human behavioral ecology. *Evolutionary Anthropology*, **16**, 107–118.

Tukey, H. 1977. *Exploratory Data Analysis*. Reading, MA: Addison-Wesley.

Wegmann, D. & L. Excoffier (2010) Bayesian inference of the demographic history of chimpanzees. *Molecular Biology and Evolution*, **27**, 1425–1435.

White, N. (1994) *Archaeological Investigations at Six Sites in the Apalachicola River Valley, Northwest Florida*. A National Oceanic and Atmospheric Administration technical memorandum. Washington, DC: US Department of Commerce, National Ocean Service.

White, T. (1991) *Human Osteology*. San Diego, CA: Academic Press.

Williams, F. L. & J. W. Patterson (2010) Reconstructing the paleoecology of Taung, South Africa from low magnification of dental microwear features in fossil primates. *Palaios*, **25**, 439–448.

Zabell, S. (2001) *Ronald Aylmer Fisher*. In *Statisticians of the Centuries*, eds. C. Heyde & E. Seneta, pp. 389–397. New York, NY: Springer.

Zar, H. (1999) *Biostatistical Analysis*. Upper Saddle River, NJ: Prentice Hall.

Zeleznik, W. S. & I. M. Bennett (1991) Assumption validity in human optimal foraging. The Bari hunters of Venezuela as a test case. *Human Ecology*, **19**, 499–508.

Index